A Quantum Mechanics Primer

DANIEL T. GILLESPIE

Institute for Molecular Physics
University of Maryland

INTERNATIONAL TEXTBOOK COMPANY
An Intext *Publisher*
Scranton, Pennsylvania 18515

ISBN 0-7002-2290-1

Printed in the United States of America by The Haddon Craftsmen, Inc., Scranton, Pennsylvania.

Library of Congress Catalog Card Number: 70-117418

To my wife

Louise

Foreword

In this book I have tried to write for science majors at the junior level a self-contained, somewhat simplified but essentially "honest" exposition of the formal theory of quantum mechanics. I have not attempted to treat any of the traditional *applications* of quantum mechanics; instead, I have endeavored only to present a concise, axiomatic development of the *theory*, with a view to bringing out as clearly as possible the main features of its mathematical and conceptual structure.

The book grew out of a series of lectures which I gave to a class of honors sophomore physics majors at Johns Hopkins University in the Spring of 1967. That experiment in an earlier-than-usual introduction to quantum mechanics convinced me that, if one is sufficiently selective and careful, students at roughly this level can be rather quickly brought to a surprisingly good understanding of the formal structure of the quantum theory. This book represents an expansion and rearrangement of those lectures, but the aim and approach are essentially unchanged.

It must be emphasized that this book cannot and is not intended to "supplant" any of the existing texts on elementary quantum mechanics. Its purpose, rather, is to *complement* and *supplement* these texts by providing the student with a simplified but meaningful *perspective* of the theory. To successfully convey such a perspective, one must place a high premium on conciseness as well as clarity; accordingly, I have tried to avoid in this book the pursuit of results not absolutely necessary for the development of the main aspects of the theory, even though such results are quite properly pursued as a matter of course in the more leisurely, comprehensive treatments found in the standard textbooks. At the same time, the restricted treatment given in this book should, by virtue of its conciseness, contribute significantly to the student's understanding and appreciation of the more elaborate developments of the standard texts. With these points in mind, I feel that this book would be suitable for any of the following uses:

(i) A textbook for a half-semester course introducing quantum mechanics to students at the junior or senior level.

(ii) A "warm-up" text or "auxiliary" text for a standard, introductory quantum mechanics course at the senior or beginning graduate level.

(iii) A text for an outside reading course for honors undergraduates.

(iv) A self-study book for students who are now taking, or who have just taken, a standard first course in quantum mechanics.

The book is meant to be completely self-contained, given the following prerequisites: a good background in "sophomore physics" (specifically, an understanding of elementary classical mechanics, plus an awareness of the inadequacy of the classical theory, as exemplified by blackbody radiation, the Bohr atom, and the Davisson-Germer experiment); a fair understanding of elementary calculus, through the improper integral and integration-by-parts; a good grasp of and feeling for the algebra of vectors in three dimensions, through the dot product; and finally, at least a nodding acquaintance with the notion of a complex number.

Because the aim and approach of this book are quite different from the standard texts on elementary quantum mechanics, a separate "Teacher's Preface" follows these remarks. There I have tried to explain very briefly the plan and organization of the book for the benefit of those who already have some knowledge of the subject.

I am happy to acknowledge the help of a number of people who have in various ways contributed to the making of this book. First of all, I owe a debt of gratitude to Professor F. Mandl, through whose extraordinarily well-written book *Quantum Mechanics* [Butterworths Publications, Ltd., London] I first learned much of what I know about this subject. The general approach of my book follows in many respects the approach of Mandl's book, and if my book can accomplish on its level something of what Mandl's does on a more advanced level, then I will be quite pleased. I want to thank my 1966-67 General Physics class at Johns Hopkins University for so willingly submitting themselves to that novel experiment; without their genuine interest in and reaction to my original lectures, this book surely could not have been written. It is a particular pleasure to thank Professor Brian R. Judd of Johns Hopkins University, not only for his graciously consenting to read two drafts of the manuscript, but also for his advice and encouragement. I am very grateful to my good friends Professor Joseph D. Sneed of Stanford University, Dr. Gary A. Prinz of NRL, Professor James S. Marsh of the University of West Florida, Dr. Edward J. Moses of Vanderbilt University, and Professor Ed S. Dorman of Western Kentucky University, for giving me their carefully considered opinions on various portions of the book. In thanking all the aforementioned individuals, I do not want to give the impression that my book necessarily mirrors in every detail their own respective views of quantum mechanics; indeed, the fact that this subject often means such different things to different people is to me one of its most fascinating and enjoyable attributes.

DANIEL T. GILLESPIE

College Park, Maryland
April, 1970

Teacher's Preface

The general plan of the book is fairly well conveyed through the Table of Contents: Chapter 1 orients the student; Chapter 2 develops the required mathematics; Chapter 3 reviews briefly the salient features of elementary classical mechanics (and in the process gives a simplified derivation and discussion of Hamilton's equations); and the remaining three-quarters of the book is given over to Chapter 4 for the development of quantum mechanics.

The quantum theory is developed for a nonrelativistic system with one degree of freedom via the position representation of the Schrödinger picture—i.e., wave mechanics in one dimension. However, every attempt is made to do this under the guise of the general theory of Dirac; more specifically, even though the Dirac "bra-ket" notation is *not* used, the central theme of the development is that quantum mechanics deals, on the mathematical level, with vectors and operators in a Hilbert space. Accordingly, $\Psi_t(x)$ is consistently referred to as the state *vector* of the system, which can be "expanded as a linear combination of the eigenvectors $\{\alpha_n(x)\}$ of the observable operator $\hat{\mathbf{A}}$, $\Psi_t(x) = \Sigma_n\ (\alpha_n,\Psi_t)\alpha_n(x)$, inasmuch as these eigenvectors form an orthonormal basis in the infinite-dimensional Hilbert space." Such an approach should convey to the student a much deeper understanding of the theory than would be conveyed by a straight "wave-function" approach, but it clearly requires a simplified yet meaningful exposition of the mathematics of the Hilbert space. This is given in Secs. 2-2, 2-3, and 2-4. The key to this exposition is in Sec. 2-3, where the concept of a Hilbert space vector is developed in close analogy with the familiar notion of a vector in three dimensions. The import of this section is summarized in the table on page 23, and warrants further comment.

The main vector analogy is actually drawn between a directed line segment v, and a complex function ψ of a real variable x; it is *not* drawn, at least initially, between their n-tuple components $\{v_i\}$ and

$\{\psi(x)\}$. In other words, in Sec. 2-3 we do not regard the x in $\psi(x)$ as a continuous component index, but rather as merely the argument of the function ψ, which itself is the abstract Hilbert space vector. This is done in order to emphasize the important fact that a vector in Hilbert space exists independently of any particular basis representation, even as does a vector in three dimensions. By drawing an analogy between the inner-product *definitions*, $\mathbf{v}_1 \cdot \mathbf{v}_2 \equiv v_1 v_2 \cos\theta_{12}$ and $(\psi_1, \psi_2) \equiv \int \psi_1^*(x)\psi_2(x)dx$, which are viewed merely as ways of deriving a scalar from two vectors such that certain rules are obeyed, one can then pass to any particular n-tuple representation of the vectors $\mathbf{v}$ and $\psi(x)$ through the sets of inner products $\{\mathbf{e}_n \cdot \mathbf{v}\}$ and $\{(\epsilon_n, \psi)\}$. After the Dirac delta function is discussed in Sec. 4-6b, it is shown that a Hilbert space vector "represents itself" with respect to the eigenbasis vectors of the position operator—i.e., that $\{(\delta_\nu, \psi)\} = \{\psi(\nu)\}$. But this fact is regarded as merely an "interesting result," and not as a key feature of the Hilbert space concept. This point of view is not only consistent, but has the important pedagogical advantage of removing the conceptual narrowness of working in the position representation, thereby bringing the student much closer to the real spirit of Dirac's coordinate-free approach.

In order to keep the development of the quantum theory as uncomplicated as possible, three simplifying restrictions are imposed: (i) the system has only one degree of freedom, (ii) operator eigenvalues are entirely discretely distributed, and (iii) operator eigenvalues are nondegenerate. However, in the final sections of the book (Secs. 4-6a, 4-6b, and 4-6c), brief discussions are given showing how the theory is modified when each of these three restrictions is removed. The Dirac delta function in particular is sidestepped as much as possible until Sec. 4-6b, and even there is discussed only briefly.

The method of presentation in Chapter 4 is postulative-deductive: the game is to derive, develop and synthesize the implications of six fundamental postulates into a coherent picture of quantum mechanics. These postulates are found on the following pages:

Throughout the development it is emphasized that the chief difference between classical and quantum mechanics lies not in their respective schemes for the time evolution of a system, but rather in their respective concepts of "state," "observables," and "measurement." The first three sections of Chapter 4, encompassing Postulates 1 through 4, are devoted to a careful elucidation of these concepts. The treatment of these matters adheres to the orthodox "Copenhagen" interpretation of quantum mechanics, inasmuch as it seems to have won the acceptance of most physicists today. The essential thesis of the Copenhagen interpretation is that quantum mechanics provides a complete and objective description of the dynamics of a single system, and in particular that it is *not* just a description of our state of knowledge of these dynamics, *nor* is it *merely* a description of the average dynamics of a statistical ensemble of systems. The Copenhagen interpretation implies some rather radical conclusions with regard to the concepts of "state," "observables," and "measurement," and by and large these conclusions are developed openly and taken seriously (as in, for example, the discussion in Sec. 4-3b of the "value of an observable," and the discussion in Sec. 4-5b of the "wave-particle duality"). In adhering to the Copenhagen interpretation, it is not intended to imply that this interpretation is necessarily better or more correct than any other; but the fact is that if one wishes to "understand" quantum mechanics at all, then one cannot avoid adopting, if only tentatively, *some* particular point of view, and it seems reasonable at this stage to adopt the most commonly accepted one.

The discussion of the theory of measurement in Sec. 4-3 is based on a rather idealistic or simplistic definition of a measurement —that it is an in-principle well-defined physical operation, which, when performed on a system, yields a single, errorless, real number. The concepts of the expectation value and uncertainty are developed with special care, the groundwork for this treatment having been laid earlier in Sec. 2-1 by discussing a simple "random drawing" experiment. A careful discussion of the problem of compatibility is given, which culminates in stating and proving the Compatibility Theorem and the Heisenberg Uncertainty Principle.

In Sec. 4-4, the general problem of the time evolution of the quantum state is taken up (Postulate 5). Emphasis is laid on the similarities between the time evolutions of the quantum and classical states. Also stressed is the intimate connection between the energy of a system and its temporal development, as is exemplified by: the time-evolution operator, the time-energy uncertainty principle, the requirement for an observable to be a constant of the motion, and the important role of the stationary states. This book does not take the approach, as some do, of identifying the energy operator with the operator $i\hbar\partial/\partial t$; it is felt that this is inconsistent in a nonrelativistic theory, where the Hilbert space vectors are not functions of time.

The introduction of the *specific* observable operators $\hat{X} = x$ and $\hat{P} = -i\hbar\partial/\partial x$ is finally made in Sec. 4-5 (Postulate 6). Once these two operators are introduced, the two Schrödinger equations, the position-momentum uncertainty relation, the concepts of position probability density and position probability current, and the Ehrenfest equations, are all *derived* from the more general formulation of the preceding four sections. Also prominently featured in this section are a discussion of the wave-particle duality, a discussion of the classical limit of quantum mechanics, and a treatment of the infinite-square well (the only "real problem" that is considered in the entire book).

The book concludes with Sec. 4-6, which attempts to give the student a rough idea of what happens when the previously mentioned simplifying restrictions are removed. This section is intended chiefly to smooth the way to the more detailed and comprehensive treatments of the standard texts.

A total of 73 exercises are closely interwoven with the text. Most of these exercises are short and easy; they are mainly intended to give the student a greater sense of active participation in the development of the theory.

The concise formulation of the Compatibility Theorem is taken from the book of F. Mandl,† which also inspired the analogy between $\mathcal{E}_3$ and $\mathcal{H}$, as well as the overall approach of the fundamental postulates. The development of the time-energy uncertainty relation and the discussion of the classical interpretation of Ehrenfest's equations follow the treatments given in the text of A. Messiah.‡ The author gratefully acknowledges the special help received from these two "standard" textbooks, and commends both to the reader who is interested in pursuing quantum mechanics beyond the confines of this book.

†F. Mandl, *Quantum Mechanics*, Butterworths Publications, Ltd., London, 1957.

‡Albert Messiah, *Quantum Mechanics* (2 vols.), North-Holland Publishing Co., Amsterdam, 1962.

Contents

CHAPTER

1

Introduction

Toward the end of the nineteenth century it seemed quite apparent to all physicists that the general concepts of what we now call "classical physics" were adequate to describe all physical phenomena. Classical mechanics, first formulated by Isaac Newton in the late seventeenth century, had by this time reached full bloom, and evidently provided a completely valid framework for the treatment of the dynamics of material bodies. Complementing classical mechanics was classical electrodynamics, finalized by James Clerk Maxwell in the latter half of the nineteenth century, which described all the properties of the electromagnetic field, and which in particular gave an intelligible account of the wave nature of light.

During the first quarter of the twentieth century, as physicists turned from their successful treatment of the macroscopic world to an examination of the microscopic world, a number of unexpected difficulties arose. Broadly speaking, these difficulties fell into two general categories.

First was the discovery of instances in nature in which certain physical variables assumed only *quantized* or *discrete* values, in contrast to the continuum of values expected on the basis of classical physics. For example, in order to explain the observed intensity spectrum of electromagnetic radiation inside a constant-temperature cavity (so-called "black-body radiation"), Max Planck in 1900 found it necessary to permit each atomic oscillator in the walls of the cavity to radiate energy only in the discrete amounts

$$h\nu,\ 2h\nu,\ 3h\nu,\ \ldots$$

Here, ν is the intrinsic frequency of the radiating oscillator (the cavity walls were assumed to contain oscillators of all frequencies), and h is a universal constant, now called *Planck's constant*, with the value

$$h = 6.625 \times 10^{-34} \text{ joule} \cdot \text{sec} \tag{1-1}$$

As another example, in order to account for the spectrum of radiation emitted by excited hydrogen atoms, Niels Bohr in 1913 found it necessary to permit the angular momentum of the orbital electrons to have only the discrete values

$$h/2\pi,\ 2h/2\pi,\ 3h/2\pi,\ \ldots$$

There were several other instances of such "quantum effects" uncovered in the early part of the twentieth century. In each case, the quantization of the appropriate variable amounted to an *ad hoc* hypothesis, and was without precedent in earlier applications of classical physics.

The second category of difficulties which beset classical physics, as applied to the microscopic world, concerned the distinction between *waves* and *particles*. By 1900 it was generally believed that light was a wave, while the electron was a particle. However, concerning the nature of *light*, Albert Einstein in 1905 put forth his theory of the photoelectric effect, which indicated that a light beam of frequency ν behaves as though it were a collection of *particles*, each with an energy

$$\epsilon = h\nu$$

Einstein's hypothesis was a bold extrapolation of Planck's theory of blackbody radiation, but it was subsequently borne out in great detail by precise experimental investigations of the photoelectric effect; it received further dramatic support in 1923 when A. H. Compton showed that these light particles, called "photons," could actually be bounced off electrons according to the usual rules of classical mechanics. Meanwhile, concerning the nature of the *electron*, C. Davisson and L. Germer showed in 1927 that, by scattering a beam of electrons off a crystalline lattice of atoms, one could obtain diffraction patterns virtually identical to those which result from the crystal-scattering of X-rays. In fact, they showed that a beam of electrons of momentum p produced a diffraction pattern characteristic of a *wave* with wavelength

$$\lambda = h/p \tag{1-2}$$

in exact agreement with the conjecture made three years earlier by L. de Broglie. In short, light was found to behave sometimes as a particle and sometimes as a wave, and the electron was found to behave sometimes as a particle and sometimes as a wave! These results evidently implied some sort of "wave-particle duality" in nature which was quite unintelligible in terms of purely classical concepts.

It gradually became apparent during the first part of the twentieth century that these two difficulties—namely the quantization of physical variables and the wave-particle paradox—bore the totally unexpected message that the *microscopic* world was simply not intelligible in the context of classical physics, and that a radically different approach was needed. As it happened, such a new approach was not long in coming: by 1930, through the efforts of W. Heisenberg, I. Schrödinger, M. Born, N. Bohr, P. A. M. Dirac, and many other physicists, a bold new system of mechanics called "quantum mechanics" had been devised.

The basic tenets of quantum mechanics are in many respects quite foreign to the concepts and attitudes of classical physics—so much so that there were and still are many eminent physicists who find some of these tenets philosophically unsatisfactory. Certainly it would be presumptuous to assert absolutely that quantum mechanics, as currently formulated, is the only or even the best possible way of understanding physical phenomena. However, there is no denying the fact that quantum mechanics, in its present form, has been amazingly successful from an "operational" point of view; that is, its predictions, no matter how unusual, have always been very much in accord with experimental observations. This, of course, is the reason for the acceptance of modern quantum theory by the overwhelming majority of physicists today.

It is our intent in this book to give a concise, simplified account of the main theoretical structure of quantum mechanics. To this end, we begin in Chapter 2 by presenting the "mathematical language" of quantum mechanics, assuming on the part of the reader mainly a reasonable grasp of elementary calculus. In Chapter 3 we review briefly the essential features of classical mechanics, in order that we may be able to readily compare and contrast the new with the old. In Chapter 4 we develop, under several simplifying restrictions, the basic formalism of quantum mechanics. Our method of presentation will be essentially "postulative-deductive"; that is, we shall lay out a number of postulates, and we shall try to derive, develop, and synthesize the implications of these postulates into a reasonably coherent theoretical framework. We shall emphasize neither the historical evolution of quantum mechanics, nor the applications of the theory to the solutions of various types of problems. Rather, we shall be chiefly concerned with understanding the *structure* and *spirit* of the theory itself. In particular, we shall try to see how quantum mechanics manages to subsume under a single, self-consistent point of view, the "common sense" of *macroscopic* physics along with the "obvious paradoxes" of *microscopic* physics. Following our development of the general theory, we shall consider briefly its application to one simple idealized physical system, and this we do merely in order to illustrate various aspects of the theory. We conclude with a short discussion of how the removal of some of the simplifying restrictions which we imposed upon our development of quantum mechanics, can be expected to affect the overall theory.

There are a liberal number of exercises sprinkled in with the text. The majority of these exercises are not of the "problem-solving" variety; rather, their solutions tend to form an integral part of the text. For this reason, most of the exercises cannot be skipped over without severely impairing the entire presentation.

The fact that this book largely ignores the many applications of quantum mechanics should not be taken to imply that these applications are irrelevant to the problem of understanding the theory. It is true that the *theory* of quantum mechanics provides a conceptual setting for the various applications, thereby interrelating these applications in a logically satisfying way; however, it is also true that the specific *applications* of quantum mechanics provide concrete examples of the highly abstract theory, thereby rendering the theory intelligible and retainable. Thus a *real* understanding of quantum mechanics can come only after both its theory *and* its applications have been thoroughly studied, each in the light of the other. Since this little book is confined to an elementary presentation of the theory only, it obviously cannot carry the reader all the way to the level of a full-fledged "quantum mechanician." However, it is hoped that the use of this book as an introduction or supplement to the more conventional textbooks and courses in quantum mechanics will help to steady and lengthen the reader's "first steps" in this journey.

CHAPTER

2

The Mathematical Language of Quantum Mechanics

Classical mechanics is formulated in terms of the mathematical language of differential and integral calculus. For example, velocity and acceleration are defined in terms of the derivative, work and impulse are defined in terms of the integral, and the conservation principles of energy and momentum find their rigorous justifications in certain elementary theorems of calculus. Quantum mechanics, too, has a mathematical language—a language that involves not only calculus but also several other branches of mathematics. In this chapter we present, in as concise and elementary a way as we can, those mathematical concepts (other than calculus) which are essential to a meaningful understanding of quantum mechanics. The necessity for achieving a reasonable degree of fluency in this mathematical language is even greater in the case of quantum mechanics than classical mechanics; for quantum theory unfortunately does not readily lend itself to nonmathematical clarifications in terms of notions familiar to us from everyday experience. The reader is therefore urged to gain a full understanding of the material presented in this chapter before proceeding to the following chapters.

2-1 PROBABILITY AND STATISTICS

In order to develop several concepts of probability theory that we shall need in our discussion of quantum mechanics, let us imagine that we have a box which contains N balls, each marked with some number which we denote generically by v. In general, the same v-number may appear on more than one ball, and we let n_k be the number of balls on which there appears the particular v-number v_k. The box of balls is therefore described by the two sets of numbers

$v_1, v_2, v_3, \ldots$, and $n_1, n_2, n_3, \ldots$. Evidently, the integers $\{n_k\}$ satisfy $\Sigma_k n_k = N$.†

Suppose we select a ball at random from the box; what is the *probability* p_k that the selected ball will show the value v_k? Since out of N possible selections, n_k of these would yield the v-number v_k, we conclude that

$$p_k = \frac{n_k}{N} \tag{2-1}$$

Thus if $n_k = 0$ it would be *impossible* to select a ball showing v_k, and we would have $p_k = 0$; on the other hand, if $n_k = N$ it would be an *absolute certainty* that the selected ball would show v_k, and we would have $p_k = 1$. In general, the numbers $\{p_k\}$ satisfy the conditions

$$0 \leqq p_k \leqq 1 \text{ for all } k \tag{2-2a}$$

and

$$\sum_k p_k = 1 \tag{2-2b}$$

Exercise 1. Prove Eqs. (2-2).

Let us calculate the probability that a single random selection from the box will yield a ball showing *either* v_k or v_j. Since out of N possible selections, a total of $(n_k + n_j)$ would yield one of these v-numbers, we conclude that

$$p\,(\text{either } v_k \text{ or } v_j) = \frac{n_k + n_j}{N} = p_k + p_j \tag{2-3a}$$

In light of this result, we may view Eq. (2-2b) as simply stating that it is an absolute certainty that a randomly selected ball will show *some* v_k number.

Suppose we now make *two* random selections, taking care to return to the box the first ball selected before making the second selection (thus, it is possible to pick the same ball both times). What is the probability that the first ball will show the value v_k and the second ball then show the value v_j? There are n_k ways to select a v_k-ball, and for each of these ways there are n_j ways to select a v_j-ball; thus, there are a total of $n_k \cdot n_j$ ways to select first a v_k-ball and then a v_j-ball. However, there are N possible selections for the first ball, and for each of these there are N possible second selections; thus,

†For the sake of brevity, we shall often denote a *set* of entities $a_1, a_2, a_3, \ldots$ by $\{a_i\}$.

there are a total of $N \cdot N$ possible double selections. We conclude then that

$$p\,(\text{first } v_k\text{, then } v_j) = \frac{n_k \cdot n_j}{N \cdot N} = p_k \cdot p_j \tag{2-3b}$$

Equations (2-3) form the basis for almost all considerations involving probability theory.

Exercise 2. In the situation we have been discussing, suppose the box contains $N = 50$ balls, each bearing some integer between 1 and 8; specifically, letting n_k be the number of balls showing the value $v_k = k$, suppose that $n_1 = 3$, $n_2 = 2$, $n_3 = 5$, $n_4 = 8$, $n_5 = 13$, $n_6 = 9$, $n_7 = 6$ and $n_8 = 4$. Use the probability concepts developed above to calculate the probability that the numbers found on two random samplings will sum to 5. [*Ans.:* 68/2500]

Suppose now that we subject our box of N balls to M samplings; that is, we select a ball at random from the box, record its v-number and return it to the box a total of M times. We denote by $v^{(i)}$ the v-value recorded on the ith sampling, and we make the following two definitions: The *mean* or *average* of the v-values recorded is

$$\langle v \rangle \equiv \frac{\sum_{i=1}^{M} v^{(i)}}{M} \tag{2-4}$$

and the *root-mean-square* (or rms) *deviation* of these values is

$$\Delta v \equiv \sqrt{\frac{\sum_{i=1}^{M} (v^{(i)} - \langle v \rangle)^2}{M}} \tag{2-5}$$

The definition of $\langle v \rangle$ is undoubtedly familiar and needs little comment. It describes the way in which we would ordinarily compute the "best value" of a series of measurements, or the "average grade" on a class quiz. The latter analogy is actually more appropriate to our discussion here, since we evidently do *not* wish to imply that $\langle v \rangle$ has some truth or legitimacy beyond that of any of the individual $v^{(i)}$-values.

Less familiar, perhaps, than the definition of the mean value $\langle v \rangle$ is the rms deviation Δv. We see that to compute this quantity, we first calculate the deviation from the mean, $v^{(i)} - \langle v \rangle$, of each v-number obtained; we next compute the average of the squares of these deviations (the squares are taken to keep the positive and negative deviations from canceling each other); and finally, to counteract to some extent the squaring, we take the square root of this average. Thus Δv is the square *root* of the *mean* of the *squares* of the *deviations* of the $v^{(i)}$-values from $\langle v \rangle$. This quantity might also be called the *rms dispersion*, since it evidently measures the extent to which the $v^{(i)}$-values are "dispersed" about $\langle v \rangle$. Of course, this is not the only quantity which can be calculated to measure this dispersion; for example, we could compute instead the average of the absolute values of the deviations, $|v^{(i)} - \langle v \rangle|$. However, the quantity in Eq. (2-5) has the advantage that it can be written in another often useful form. Specifically, we see from Eq. (2-5) that

$$(\Delta v)^2 = \frac{\sum_{i=1}^{M} [v^{(i)2} - 2\langle v \rangle v^{(i)} + \langle v \rangle^2]}{M}$$

$$= \frac{\sum_{i=1}^{M} v^{(i)2}}{M} - 2\langle v \rangle \frac{\sum_{i=1}^{M} v^{(i)}}{M} + \frac{M\langle v \rangle^2}{M}$$

$$= \langle v^2 \rangle - 2\langle v \rangle\langle v \rangle + \langle v \rangle^2$$

Therefore,

$$\Delta v = \sqrt{\langle v^2 \rangle - \langle v \rangle^2} \tag{2-6}$$

In words, the rms deviation of the $v^{(i)}$-values is equal to the square root of the difference between the *average of the square* and the *square of the average*. It is to be noted that these two quantities are *not* in general equal; indeed, a comparison of Eqs. (2-5) and (2-6) reveals that $\langle v^2 \rangle = \langle v \rangle^2$ *only if every* $v^{(i)}$-value coincides with $\langle v \rangle$. Equation (2-6) tells us that the extent to which $\langle v^2 \rangle$ and $\langle v \rangle^2$ differ provides us with a direct measure of the dispersion in the $v^{(i)}$-values.

If we have a knowledge of the two sets of numbers $\{v_k\}$ and $\{n_k\}$, or equivalently $\{v_k\}$ and $\{p_k\}$, it would seem that we ought to be able to *predict* approximately what values would be obtained for $\langle v \rangle$ and Δv. The key to making such a prediction is the following assumption: since n_k of the N balls have the number v_k, then in M random samplings of these balls we ought to obtain the value v_k approximately m_k times, where $m_k/M = n_k/N$. Thus, using Eq. (2-1), the approximate number of times the value v_k should appear in the set of values $v^{(1)}, v^{(2)}, \ldots, v^{(M)}$ is

$$m_k = \frac{n_k}{N} M = p_k M$$

With this, the sum in Eq. (2-4) can be written

$$\sum_{i=1}^{M} v^{(i)} = \sum_k m_k v_k = \sum_k (p_k M) v_k$$

and Eq. (2-4) becomes

$$\langle v \rangle = \sum_k p_k v_k \tag{2-7}$$

Equation (2-7) expresses $\langle v \rangle$ as a "weighted sum" of the possible v_k-values; the weight assigned to any particular value v_k is just the probability of its occurrence, p_k. It should be remarked that this value for $\langle v \rangle$ is the "theoretically expected" value; the "experimental" value in Eq. (2-4) will generally differ somewhat from this theoretical value owing to the randomness involved. However, in the limit of very many experimental samplings ($M \longrightarrow \infty$), the value in Eq. (2-4) may be expected to get arbitrarily close to the value in Eq. (2-7).

Equation (2-7) may be generalized quite easily, as the following exercise shows.

Exercise 3. Let f be a given function of v, and let this function be evaluated for each of the $v^{(i)}$-values. Prove that the average or mean of the resulting set of $f(v^{(i)})$-values is

$$\langle f(v) \rangle = \sum_k p_k f(v_k) \tag{2-8}$$

[Note that by putting $f(v) = v$ in Eq. (2-8), we obtain Eq. (2-7).]

By putting $f(v) = v^2$ in Eq. (2-8), we see that

$$\langle v^2 \rangle = \sum_k p_k v_k^{\,2}$$

Using this and Eq. (2-7), we may thus write Eq. (2-6) as

$$\Delta v = \sqrt{\left(\sum_k p_k v_k^{\,2}\right) - \left(\sum_k p_k v_k\right)^2} \tag{2-9}$$

We now observe that Eqs. (2-7) and (2-9) express the two basic quantities $\langle v \rangle$ and Δv wholly in terms of the numbers $\{v_k\}$ and $\{p_k\}$. Thus, given a set of values $v_1, v_2, \ldots$ distributed with probabilities p_1, $p_2, \ldots$, Eqs. (2-7) and (2-9) allow us to calculate the theoretically expected *mean* and *rms deviation* to be obtained in any random sampling of these v-values.

Exercise 4. Consider the collection of numbered balls described in Exercise 2.

(a) Calculate $\langle v \rangle$ and Δv. [*Ans.*: $\langle v \rangle = 4.94$ and $\Delta v = 1.8$]

(b) Sketch a "frequency bar-graph" of the expected results of $M = 100$ samplings [i.e., lay out the values v_k on the horizontal axis, and construct vertical "bars" to indicate the number of times each v_k-value should be obtained.] Show on the graph by means of a vertical line the value $\langle v \rangle$. Also, draw a horizontal line of length $2\Delta v$ in such a way that it indicates roughly the "spread" or "dispersion" of the v-values about $\langle v \rangle$.

The foregoing exercise illustrates the overall significance of $\langle v \rangle$ and Δv. Certainly a *complete* description of the expected results of a "multiple sampling" experiment requires the specification of *all* the numbers (v_1, p_1), (v_2, p_2), (v_3, p_3), ... However, if we are asked to describe the results with *only two* numbers, we would evidently do well to state the values of $\langle v \rangle$ and Δv: $\langle v \rangle$ is essentially a "collective value" for the set of v-numbers, while Δv (or the smallness thereof) provides a quantitative measure of the degree to which it is actually meaningful to so characterize the *set* of v-values by a *single* value. These ideas will play a very important role in understanding certain basic concepts in quantum mechanics.

2-2 COMPLEX NUMBERS

An understanding of quantum mechanics requires some knowledge of a few elementary properties of complex numbers. For our purposes, we may define a complex number c as a quantity which can be written

$$c \equiv a + ib \tag{2-10a}$$

Here a and b are ordinary real numbers, while the "number" i satisfies

$$i^2 \equiv -1 \quad \text{or} \quad i \equiv \sqrt{-1} \tag{2-10b}$$

The real number a is called the "real part" of c, and the real number b is called the "imaginary part" of c:

$$a \equiv \text{Re}c \qquad b \equiv \text{Im}c \tag{2-10c}$$

If $b = 0$, then c is said to be a "pure real" number; if $a = 0$, then c is said to be a "pure imaginary" number. We write $c = 0$ if and only if $a = b = 0$.

Complex numbers can be added and multiplied. The rules for carrying out these operations are the same as for ordinary real numbers, but taking account of Eq. (2-10b). Thus if $c_1 = a_1 + ib_1$ and $c_2 = a_2 + ib_2$, then the *sum* of c_1 and c_2 is defined to be the complex number

$$c_1 + c_2 \equiv (a_1 + ib_1) + (a_2 + ib_2) = (a_1 + a_2) + i(b_1 + b_2) \tag{2-11a}$$

and the *product* of c_1 times c_2 is defined to be the complex number

$$c_1 \cdot c_2 \equiv (a_1 + ib_1) \cdot (a_2 + ib_2) = (a_1 a_2 - b_1 b_2) + i(a_1 b_2 + b_1 a_2) \tag{2-11b}$$

If c is written in the form of Eq. (2-10a), then the *complex conjugate* of c is defined to be the complex number

$$c^* \equiv a - ib \tag{2-12}$$

Exercise 5. Prove the following properties of the complex conjugate:

(a) $$\text{Re}c = \frac{c + c^*}{2} \qquad \text{Im}c = \frac{c - c^*}{2i} \tag{2-13}$$

(b) c is a pure *real* number if and only if $c^* = c$. c is a pure *imaginary* number if and only if $c^* = -c$.

(c) $$c^{**} = c \tag{2-14a}$$

$$(c_1 + c_2)^* = c_1^* + c_2^* \tag{2-14b}$$

$$(c_1 c_2)^* = c_1^* c_2^* \tag{2-14c}$$

The *square modulus* of c is denoted by $|c|^2$ and is defined to be the product of c times its complex conjugate:

$$|c|^2 \equiv c^* c \tag{2-15a}$$

The *modulus* of a complex number is just the positive square root of its square modulus:

$$|c| \equiv +\sqrt{c^* c} \tag{2-15b}$$

Exercise 6. Prove the following properties of the modulus:

(a) $$|c|^2 = (\text{Re}c)^2 + (\text{Im}c)^2 \quad \text{or} \quad |c| = \sqrt{(\text{Re}c)^2 + (\text{Im}c)^2} \tag{2-16}$$

(b) $$|c| \geqq |\text{Re}c| \quad \text{and} \quad |c| \geqq |\text{Im}c| \tag{2-17}$$

(c) $$|c_1 c_2| = |c_1|\,|c_2| \tag{2-18a}$$

$$|c_1 + c_2| \leqq |c_1| + |c_2| \tag{2-18b}$$

[*Hint:* Write $c_1 = a_1 + ib_1$ and $c_2 = a_2 + ib_2$ for part (c).]

The *square modulus* of c should not be confused with the *square* of c. If $c = a + ib$, then using Eq. (2-11b), we have

$$|c|^2 = c^*c = a^2 + b^2$$

whereas

$$c^2 = cc = (a^2 - b^2) + i(2ab)$$

Clearly, *the square modulus is always a nonnegative real number*, while the square is in general complex. In fact, the distinction between the square and the square modulus disappears if and only if c is pure real; in this case, the modulus becomes the absolute value.

It should be remarked that the complex number system can be set up without making use of the "number" $\sqrt{-1}$. This is done by initially *defining* a complex number c to be an ordered pair of real numbers (a,b), and then setting forth appropriate rules for algebraically manipulating these ordered pairs. For example, if $c_1 = (a_1, b_1)$ and $c_2 = (a_2, b_2)$, then the sum $c_1 + c_2$ would be *defined* as the ordered pair $(a_1 + a_2, b_1 + b_2)$, and the product $c_1 \cdot c_2$ would be *defined* as the ordered pair $(a_1 a_2 - b_1 b_2, a_1 b_2 + a_2 b_1)$. The symbol i can then be introduced as merely an *ad hoc* device to simplify these rules; that is, by writing the ordered pair (a,b) as $a + ib$, we can manipulate complex numbers using the familiar rules of the algebra of *real* numbers, with the additional rule that i^2 is always to be replaced by -1. Although we shall always write $a + ib$ instead of (a,b), the reader should try to adopt this essentially "algebraic" attitude toward complex numbers instead of the popular "mystical" attitude, which is overly concerned with the square roots of negative real numbers.

In exact analogy with the foregoing, we can define a *complex function* ψ of a *real variable* x to be a function of the form

$$\psi(x) = u(x) + iv(x) \tag{2-19}$$

where $u(x)$ and $v(x)$ are ordinary real functions of the real variable x. All the preceding equations concerning the complex number c hold for the complex function $\psi(x)$, provided that we replace Rec by Re$\psi(x) = u(x)$, and Imc by Im$\psi(x) = v(x)$. Thus, for example, the complex conjugate of $\psi(x)$ is $\psi^*(x) = u(x) - iv(x)$, and the square modulus of $\psi(x)$ is $|\psi(x)|^2 = u^2(x) + v^2(x)$.

The complex function $\psi(x)$ can be differentiated and integrated with respect to its argument x. The rules for carrying out these two operations are just what one would expect:

$$\frac{d}{dx}\psi(x) \equiv \frac{d}{dx}u(x) + i\frac{d}{dx}v(x)$$

and

$$\int_a^b \psi(x)dx \equiv \int_a^b u(x)dx + i \int_a^b v(x)dx$$

We note that the derivative of $\psi(x)$ is a complex function of x, whereas the definite integral of $\psi(x)$ is a complex number, in exact analogy with the situation for real functions of x.

It should perhaps be mentioned that it is also possible to define complex functions ψ of a *complex variable* $z = x + iy$. The situation with respect to differentiation and integration then becomes rather involved. However, in this book we shall require only a knowledge of the comparatively simple properties of complex functions of a real variable, as outlined above.

Exercise 7. The complex function e^{ikx} (k real) is *defined* by

$$e^{ikx} \equiv \cos kx + i \sin kx \tag{2-20a}$$

Prove from this definition that e^{ikx} has the following properties:

$$(e^{ikx})^* = e^{-ikx} \tag{2-20b}$$

$$e^{ik_1x} \cdot e^{ik_2x} = e^{i(k_1+k_2)x} \tag{2-20c}$$

$$|e^{ikx}|^2 = 1 \tag{2-20d}$$

$$\frac{d}{dx} e^{ikx} = (ik)\, e^{ikx} \tag{2-20e}$$

$$\int e^{ikx}\, dx = \frac{1}{ik} e^{ikx} + C \tag{2-20f}$$

[We sometimes will write exp (ikx) instead of e^{ikx}.]

2-3 HILBERT SPACE VECTORS

The language of quantum mechanics is mainly the language of a branch of mathematics called "vector spaces." The reader is assumed to be familiar with the elementary properties of "ordinary vectors" in three-dimensional Euclidean space (mathematicians denote this space symbolically by $\mathcal{E}_3$). Actually, the notion of a vector space is much more general than this. In fact, quantum mechanics is formulated in terms of an *infinite*-dimensional vector space called a "Hilbert space" (denoted symbolically by $\mathcal{H}$); in $\mathcal{H}$ the "vectors" are *not* directed line segments, as in $\mathcal{E}_3$, but rather are *complex functions of real variables.* A complete development of the mathematics of the Hilbert space is beyond our reach here; however, at the expense of a little mathematical rigor and generality, we shall find it possible to come to a fairly good understanding of the Hilbert space by drawing suitable analogies with the simpler, more familiar properties of $\mathcal{E}_3$. To this end, we shall begin our discussion of vectors in $\mathcal{H}$ by reviewing some of the salient ideas concerning vectors in $\mathcal{E}_3$.

A "vector" in $\mathcal{E}_3$ can be defined as a *directed line segment.* Thus, a vector $\mathbf{v}$ in $\mathcal{E}_3$ possesses the properties of magnitude and direction; the magnitude of $\mathbf{v}$, written $|\mathbf{v}|$, is the length of the line segment, and the direction of $\mathbf{v}$ is the direction of travel from the "tail" of the line segment to the "head" of the line segment.

Two operations common to all vector spaces are the operations of *scalar multiplication* and *vector addition. Scalars* in $\mathcal{E}_3$ are simply the set of all *real numbers.* The multiplication of a vector $\mathbf{v}$ by a scalar r yields a new vector, written $r\mathbf{v}$, whose direction is the same as that of $\mathbf{v}$ but whose magnitude is $|r|$ times the magnitude of $\mathbf{v}$ ($|r\mathbf{v}| = |r|\,|\mathbf{v}|$); negative scalar multipliers reverse the direction. The addition of two vectors $\mathbf{v}_1$ and $\mathbf{v}_2$ yields a new vector, which is written $\mathbf{v}_1 + \mathbf{v}_2$; this vector is obtained by placing the tail of $\mathbf{v}_2$ at the head of $\mathbf{v}_1$, and then constructing the directed line segment from the tail of $\mathbf{v}_1$ to the head of $\mathbf{v}_2$. These two operations of scalar multiplication and vector addition allow us to form *linear combinations* of vectors; thus, if $\mathbf{v}_1$ and $\mathbf{v}_2$ are any two vectors in $\mathcal{E}_3$, and r_1 and r_2 are any two $\mathcal{E}_3$-scalars (i.e., real numbers), then the "linear combination"

$$\mathbf{v} = r_1\mathbf{v}_1 + r_2\mathbf{v}_2 \tag{2-21}$$

is a well-defined vector in $\mathcal{E}_3$.

Another important feature of many (but not all) vector spaces is the existence of an operation called the *inner product.* In $\mathcal{E}_3$ the inner product of two vectors $\mathbf{v}_1$ and $\mathbf{v}_2$ is written $\mathbf{v}_1 \cdot \mathbf{v}_2$, and is customarily called the "dot product" of $\mathbf{v}_1$ and $\mathbf{v}_2$. By *definition*,

$$\mathbf{v}_1 \cdot \mathbf{v}_2 \equiv |\mathbf{v}_1|\,|\mathbf{v}_2| \cos\theta_{12} \tag{2-22}$$

where θ_{12} is the angle between $\mathbf{v}_1$ and $\mathbf{v}_2$ when these two vectors are placed tail-to-tail. Geometrically, $\mathbf{v}_1 \cdot \mathbf{v}_2$, can be thought of as the product of the length of $\mathbf{v}_1$ times the projected length $|\mathbf{v}_2| \cos\theta_{12}$ of $\mathbf{v}_2$ on $\mathbf{v}_1$, or equivalently as the product of the length of $\mathbf{v}_2$ times the projected length $|\mathbf{v}_1| \cos\theta_{12}$ of $\mathbf{v}_1$ on $\mathbf{v}_2$.

It is evident from Eq. (2-22) that *the inner product of two vectors is always a scalar* (in this case, a real number). In particular, the inner product of a vector with itself, called the *norm* of the vector, is always a *nonnegative real number*:

$$\text{Norm of } \mathbf{v} \equiv \mathbf{v} \cdot \mathbf{v} = |\mathbf{v}|^2 \geqq 0 \tag{2-23}$$

It can be shown from Eq. (2-22) that the inner product in $\mathcal{E}_3$ satisfies the following relations:

$$\mathbf{v}_2 \cdot \mathbf{v}_1 = \mathbf{v}_1 \cdot \mathbf{v}_2 \tag{2-24a}$$

$$r_1 \mathbf{v}_1 \cdot r_2 \mathbf{v}_2 = r_1 r_2 \mathbf{v}_1 \cdot \mathbf{v}_2 \tag{2-24b}$$

$$(\mathbf{v}_1 + \mathbf{v}_2) \cdot (\mathbf{v}_3 + \mathbf{v}_4) = \mathbf{v}_1 \cdot \mathbf{v}_3 + \mathbf{v}_1 \cdot \mathbf{v}_4 + \mathbf{v}_2 \cdot \mathbf{v}_3 + \mathbf{v}_2 \cdot \mathbf{v}_4 \tag{2-24c}$$

$$|\mathbf{v}_1 \cdot \mathbf{v}_2| \leqq \sqrt{\mathbf{v}_1 \cdot \mathbf{v}_1}\,\sqrt{\mathbf{v}_2 \cdot \mathbf{v}_2} \tag{2-24d}$$

Exercise 8. Verify the foregoing relations.

Equation (2-24a) says that the inner product is cummutative. Equations (2-24b) and (2-24c) show how the inner product behaves with respect to the operations of scalar multiplication and vector addition. Equation (2-24d) states a very fundamental property of the inner product; this relation is often referred to as the Schwarz inequality.

Two vectors are said to be *orthogonal* if they are perpendicular to each other. Since $\cos(\pi/2) = 0$, then from Eq. (2-22) we can write

$$\mathbf{v}_1 \text{ and } \mathbf{v}_2 \text{ are } \textit{orthogonal} \text{ if and only if } \quad \mathbf{v}_1 \cdot \mathbf{v}_2 = 0 \tag{2-25}$$

Indeed, we may adopt this as our *definition* of orthogonality, if we agree to regard the null vector **0** as being orthogonal to any other vector and also to itself.

We often deal with *sets* of vectors, $\mathbf{v}_1, \mathbf{v}_2, \ldots$ or more compactly, $\{\mathbf{v}_i\}$. Concerning such sets we make the following definitions:

(i) The set $\{\mathbf{v}_i\}$ is said to be *orthonormal* if and only if each vector of the set is orthogonal to every other vector of the set, and each vector of the set has unit norm. These properties can be expressed most succinctly through the use of the so-called "Kronecker delta symbol," δ_{ij}, which is defined by

$$\delta_{ij} \equiv \begin{cases} 0 & \text{if } i \neq j \\ 1 & \text{if } i = j \end{cases} \tag{2-26}$$

Thus we have

$$\{\mathbf{v}_i\} \text{ is an } \textit{orthonormal set} \text{ if and only if } \quad \mathbf{v}_i \cdot \mathbf{v}_j = \delta_{ij} \tag{2-27}$$

(ii) The set $\{\mathbf{v}_i\}$ is said to be *complete* if and only if any vector in $\mathcal{E}_3$ can be written as a linear combination of the vectors in $\{\mathbf{v}_i\}$. In other words, $\{\mathbf{v}_i\}$ is a complete set if and only if, for any vector **v** in

$\mathcal{E}_3$, there exists at least one set of scalars $\{r_i\}$ such that $\mathbf{v} = \Sigma_i r_i \mathbf{v}_i$. It turns out that, in $\mathcal{E}_3$, any set of three or more noncoplanar vectors constitutes a complete set.

Of particular interest are those sets of vectors which are *both* orthonormal and complete; such a set is called an *orthonormal basis set*. In $\mathcal{E}_3$, there are infinitely many different orthonormal basis sets (they differ from one another by simple rotations), and all such sets have exactly *three* vectors; for this reason, $\mathcal{E}_3$ is said to be "three-dimensional." A specific orthonormal basis set in $\mathcal{E}_3$ is usually written as $(\mathbf{x},\mathbf{y},\mathbf{z})$ or $(\mathbf{i},\mathbf{j},\mathbf{k})$ or $(\mathbf{e}_1,\mathbf{e}_2,\mathbf{e}_3)$; we shall use the latter notation. Thus, we have from the orthonormality of the set $\{\mathbf{e}_i\}$,

$$\mathbf{e}_i \cdot \mathbf{e}_j = \delta_{ij} \qquad (i,j = 1,2,3) \tag{2-28}$$

Moreover, since $\{\mathbf{e}_i\}$ is complete, then given any vector $\mathbf{v}$ we can find scalars r_1, r_2 and r_3 such that

$$\mathbf{v} = \sum_{i=1}^{3} r_i \mathbf{e}_i \tag{2-29a}$$

Indeed, using Eqs. (2-24) and (2-28), we see that

$$\mathbf{e}_j \cdot \mathbf{v} = \mathbf{e}_j \cdot \left(\sum_{i=1}^{3} r_i \mathbf{e}_i\right) = \sum_{i=1}^{3} r_i(\mathbf{e}_j \cdot \mathbf{e}_i) = \sum_{i=1}^{3} r_i \delta_{ij} = r_j \tag{2-29b}$$

That is, the *expansion coefficients* or *components* r_i of $\mathbf{v}$ in the orthonormal basis $\{\mathbf{e}_i\}$ are just the scalars $\mathbf{e}_i \cdot \mathbf{v}$. Therefore, we can write Eq. (2-29a) as

$$\mathbf{v} = \sum_{i=1}^{3} (\mathbf{e}_i \cdot \mathbf{v})\mathbf{e}_i \tag{2-29c}$$

an equation which may be regarded as an identity for all $\mathbf{v}$ in $\mathcal{E}_3$, and all orthonormal basis sets $\{\mathbf{e}_i\}$. The import of Eq. (2-29c) is illustrated in Fig. 1.

Exercise 9. If two $\mathcal{E}_3$-vectors $\mathbf{a}$ and $\mathbf{b}$ have components $\{a_i\}$ and $\{b_i\}$ relative to a given basis $(\mathbf{e}_1,\mathbf{e}_2,\mathbf{e}_3)$, prove that

$$\mathbf{a} \cdot \mathbf{b} = \sum_{i=1}^{3} a_i b_i \tag{2-30a}$$

and in particular that

$$\mathbf{a} \cdot \mathbf{a} = \sum_{i=1}^{3} a_i^2 \tag{2-30b}$$

[*Hint:* Make use of Eqs. (2-24) and (2-28).]

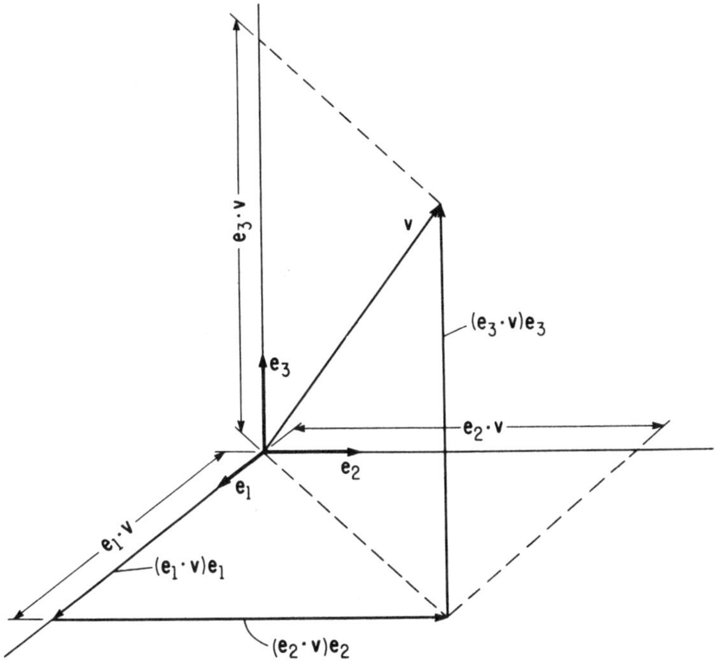

Fig. 1. Illustrating the expansion of an arbitrary vector v in an orthonormal basis set $\mathbf{e}_1$, $\mathbf{e}_2$, $\mathbf{e}_3$. The inner product $(\mathbf{e}_i \cdot \mathbf{v})$ is just the projected length or "component" of v in the direction of $\mathbf{e}_i$; therefore, v can be written as the vector sum $\mathbf{v} = \sum_{i=1}^{3} (\mathbf{e}_i \cdot \mathbf{v})\mathbf{e}_i$.

The preceding discussion of vectors in $\mathcal{E}_3$ is by no means complete, but it is extensive enough for our purposes. We shall now show how these vector concepts in $\mathcal{E}_3$ carry over into the Hilbert space $\mathcal{H}$.

We *define* a vector in $\mathcal{H}$ to be a *complex function* ψ *of a single real variable* x [see Eq. (2-19)]. In other words, a vector in $\mathcal{H}$ is a *rule of correspondence* which assigns to each real number x a complex number $\psi(x)$. To be precise, we should mention that not all such functions are truly vectors in $\mathcal{H}$, but only those functions that satisfy a certain condition; we shall state and discuss this condition later [see Eq. (2-35)].

The *scalars* in $\mathcal{H}$ are by definition the set of all *complex numbers.* This is to be contrasted with the situation in $\mathcal{E}_3$, where the scalars are the set of all *real* numbers.

The two operations of "scalar multiplication" and "vector addition" are defined in accordance with the usual rules for adding and multiplying complex quantities [see Eqs. (2-11a) and (2-11b)]. Thus, if $\psi_1(x)$ and $\psi_2(x)$ are any two vectors in $\mathcal{H}$, and c_1 and c_2 are any two $\mathcal{H}$-scalars (i.e., any two complex numbers), then the "linear combination"

$$\psi(x) = c_1\psi_1(x) + c_2\psi_2(x) \tag{2-31}$$

is a well-defined vector in $\mathcal{H}$.†

In $\mathcal{H}$ the *inner product* of two vectors $\psi_1(x)$ and $\psi_2(x)$ is written (ψ_1,ψ_2), and is *defined* by

$$(\psi_1,\psi_2) \equiv \int_{-\infty}^{\infty} \psi_1^*(x)\psi_2(x)\,dx \tag{2-32}$$

Thus to calculate (ψ_1,ψ_2), we multiply $\psi_2(x)$ by the complex conjugate of $\psi_1(x)$, and integrate the result over all values of x. This quantity is sometimes referred to as the "overlap" of $\psi_1(x)$ and $\psi_2(x)$, since it is in some loose sense a measure of the extent to which these two functions match or complement each other over the x-axis.

Exercise 10. Write $\psi_j(x) = u_j(x) + iv_j(x)$, for $j = 1$ and $j = 2$, and show explicitly that (ψ_1,ψ_2) is an $\mathcal{H}$-scalar. Note in particular that if x had *not* been integrated over in Eq. (2-32) then (ψ_1,ψ_2) would *not* have been an $\mathcal{H}$-scalar.

The fact that (ψ_1,ψ_2) is always an $\mathcal{H}$-scalar parallels the fact that $\mathbf{v}_1 \cdot \mathbf{v}_2$ is always an $\mathcal{E}_3$-scalar.

In $\mathcal{E}_3$ we defined the norm of a vector to be its inner product with itself. Analogously, in $\mathcal{H}$ we define the *norm* of the vector $\psi(x)$ to be

$$\text{Norm of } \psi(x) \equiv (\psi,\psi) = \int_{-\infty}^{\infty} \psi^*(x)\psi(x)\,dx = \int_{-\infty}^{\infty} |\psi(x)|^2\,dx \geqq 0 \tag{2-33}$$

Exercise 11. Write $\psi(x) = u(x) + iv(x)$, and show explicitly that (ψ,ψ) is a nonnegative real number. Note in particular that if one of the functions in the integrand in Eq. (2-32) had *not* been complex conjugated, then (ψ,ψ) would *not* always be a real number.

†It should be emphasized that, in denoting a Hilbert space vector by the symbol "$\psi(x)$," we are referring to the *function* ψ of the variable x, and *not* the *value* of this function at the point x.

We see then that in both $\mathcal{E}_3$ and $\mathcal{H}$, the norm of a vector is a nonnegative real number. In fact, the norm of a vector is zero only for the "null vectors," $\mathbf{v} = \mathbf{0}$ and $\psi(x) \equiv 0$.

In analogy with Eqs. (2-24), it is not difficult to show that the definition of the inner product in Eq. (2-32) implies the following properties:

$$(\psi_2, \psi_1) = (\psi_1, \psi_2)^* \tag{2-34a}$$

$$(c_1 \psi_1, c_2 \psi_2) = c_1^* c_2 (\psi_1, \psi_2) \tag{2-34b}$$

$$(\psi_1 + \psi_2, \psi_3 + \psi_4) = (\psi_1, \psi_3) + (\psi_1, \psi_4) + (\psi_2, \psi_3) + (\psi_2, \psi_4) \tag{2-34c}$$

$$|(\psi_1, \psi_2)| \leqq \sqrt{(\psi_1, \psi_1)} \sqrt{(\psi_2, \psi_2)} \tag{2-34d}$$

Exercise 12.

(a) Using the definition in Eq. (2-32), prove the first three relations listed above.

(b) Prove Eq. (2-34d) in the following way: define the vector $\psi_3(x) \equiv (\psi_1, \psi_2)\psi_1(x) - (\psi_1, \psi_1)\psi_2(x)$, and make use of the fact that the norm of $\psi_3(x)$ is nonnegative.

In this book we shall have little occasion to actually *compute* explicit inner-product integrals. However, the four *properties* of the inner product integral listed in Eqs. (2-34) will be used quite extensively, so the reader should become as familiar with them as possible. We note in particular the appearance of the complex conjugation operation in Eqs. (2-34a) and (2-34b), in contrast to Eqs. (2-24a) and (2-24b); however, even in the latter equations, the complex conjugation operation would evidently not be incorrect, but merely unnecessary. The Schwarz inequality in Eq. (2-34d) has the same form as in Eq. (2-24d), but it should be noted that $|\mathbf{v}_1 \cdot \mathbf{v}_2|$ in Eq. (2-24d) means the *absolute value* of the (possibly negative) *real* number $\mathbf{v}_1 \cdot \mathbf{v}_2$, whereas $|(\psi_1, \psi_2)|$ in Eq. (2-34d) means the *modulus* of the (in general) *complex* number (ψ_1, ψ_2).

The reader should now begin to appreciate the rationale for calling complex functions "vectors" in a vector space. From a *strictly mathematical* point of view, directed line segments may be regarded as "vectors," not because they possess the properties of magnitude and direction, but rather because we can define for directed line segments the three operations of scalar multiplication, vector addition, and vector inner multiplication, in such a way that Eqs. (2-23) and (2-24) are obeyed. It is these latter relations that actually determine the "vector character" of directed line segments, and not the concept of a directed line segment itself, nor even the specific recipes for forming the scalar product, vector sum and vector inner product, relative to directed line segments. Now we have just seen that, if we adopt certain well-defined rules for obtaining the "scalar product," "vector sum" and "inner product," relative to *complex functions of a real variable*, then we arrive at the properties expressed in Eqs. (2-33) and (2-34). These properties are essentially identical to those in Eqs. (2-23) and (2-24); consequently, we are entirely justified in regarding complex functions as "vectors" in a vector space. In particular, our definition of the inner product in Eq. (2-32), which at first sight probably seemed rather peculiar to the reader, was chosen simply because it was a way of obtaining a unique scalar from two vectors such that Eqs. (2-33) and (2-34) were satisfied. If we could conjure up a *different* set of rules for forming "linear combinations" and "inner products" of complex functions, which still satisfied all the conditions in Eqs. (2-33) and (2-34), then we would have thereby constructed *another* perfectly valid "vector space" of complex functions; however, that vector space would probably not turn out to be as relevant to the task of describing physical phenomena as our present "Hilbert space" turns out to be.

We are now in a position to state the condition alluded to earlier which a function $\psi(x)$ must satisfy in order to be a vector in $\mathcal{H}$. We admit as vectors of $\mathcal{H}$ *only* those functions $\psi(x)$ which have a *finite norm*:

$$\psi(x) \text{ is a vector of } \mathcal{H} \text{ if and only if } \quad (\psi,\psi) = \int_{-\infty}^{\infty} |\psi(x)|^2 \, dx < \infty \tag{2-35}$$

We should note that an analogous condition was implicitly imposed on $\mathcal{E}_3$-vectors, through their definition as directed line *segments* (i.e., lines of finite length). Condition (2-35) insures the following two important results:

(i) If $\psi_1(x)$ and $\psi_2(x)$ are any two vectors in $\mathcal{H}$, then the inner product (ψ_1,ψ_2) "exists" in the sense that it is a complex number with a finite modulus. To see that this is true, we note that, since $\psi_1(x)$ and $\psi_2(x)$ are $\mathcal{H}$-vectors, then $(\psi_1,\psi_1) < \infty$ and $(\psi_2,\psi_2) < \infty$ by Eq. (2-35). Our result then follows immediately from the Schwarz inequality:

$$|(\psi_1,\psi_2)| \leqq \sqrt{(\psi_1,\psi_1)}\,\sqrt{(\psi_2,\psi_2)} < \infty$$

(ii) If $\psi_1(x)$ and $\psi_2(x)$ are any two vectors in $\mathcal{H}$, then so is any linear combination $\psi(x) = c_1\psi_1(x) + c_2\psi_2(x)$. To see that this is true, we write

$$\begin{aligned}(\psi,\psi) &= (c_1\psi_1 + c_2\psi_2,\, c_1\psi_1 + c_2\psi_2)\\ &= c_1^*c_1(\psi_1,\psi_1) + c_1^*c_2(\psi_1,\psi_2) + c_2^*c_1(\psi_2,\psi_1) + c_2^*c_2(\psi_2,\psi_2)\end{aligned}$$

$$= |c_1|^2(\psi_1,\psi_1) + |c_2|^2(\psi_2,\psi_2) + c_1^*c_2(\psi_1,\psi_2) + [c_1^*c_2(\psi_1,\psi_2)]^*$$

$$= |c_1|^2(\psi_1,\psi_1) + |c_2|^2(\psi_2,\psi_2) + 2\mathrm{Re}[c_1^*c_2(\psi_1,\psi_2)]$$

so

$$(\psi,\psi) \leqq |c_1|^2(\psi_1,\psi_1) + |c_2|^2(\psi_2,\psi_2) + 2|c_1^*c_2||(\psi_1,\psi_2)|$$

Now the first two terms on the right are bounded by virtue of Eq. (2-35), and the third term is bounded by virtue of result (i) above. Consequently, $(\psi,\psi) < \infty$, and so by Eq. (2-35) the linear combination $\psi(x) = c_1\psi_1(x) + c_2\psi_2(x)$ is indeed a vector of $\mathcal{H}$.

The preceding two results mean, first, that we can be assured that the improper integrals appearing in the definition of the inner product always converge (or make sense) for Hilbert space functions, and second, that the operation of taking linear combinations of Hilbert space functions cannot produce a non-Hilbert space function.

In direct analogy with Eq. (2-25), two $\mathcal{H}$-vectors are said to be *orthogonal* if their inner product vanishes:

$$\psi_1(x) \text{ and } \psi_2(x) \text{ are } \textit{orthogonal} \text{ if and only if } (\psi_1,\psi_2) = 0 \tag{2-36}$$

A *set* of $\mathcal{H}$-vectors $\{\psi_i(x)\}$ is said to be an *orthonormal set* if and only if each vector of the set is orthogonal to every other vector of the set, and each vector of the set has unit norm. Using the Kronecker delta symbol defined in Eq. (2-26), we therefore have in analogy to Eq. (2-27),

$$\{\psi_i(x)\} \text{ is an } \textit{orthonormal set} \text{ if and only if } (\psi_i,\psi_j) = \delta_{ij} \tag{2-37}$$

A set of $\mathcal{H}$-vectors $\{\psi_i(x)\}$ is said to be a *complete set* if and only if any vector in $\mathcal{H}$ can be written as a linear combination of the vectors in $\{\psi_i(x)\}$. In other words, $\{\psi_i(x)\}$ is a complete set if and only if, for any vector $\psi(x)$ in $\mathcal{H}$, there exists at least one set of scalars $\{c_i\}$ such that $\psi(x) = \Sigma_i c_i\psi_i(x)$.

Special use will be made of sets of $\mathcal{H}$-vectors which are *both* orthonormal and complete; such a set is called an *orthonormal basis set*. In $\mathcal{H}$, as in $\mathcal{E}_3$, there are infinitely many such orthonormal basis sets. However, whereas in $\mathcal{E}_3$ all such sets contain exactly three vectors, it turns out that in $\mathcal{H}$ all orthonormal basis sets contain infinitely many vectors; for this reason $\mathcal{H}$ is said to be *infinite-dimensional*. If $\{\epsilon_i(x)\}$ is an orthonormal basis set, then we have, by the orthonormality of the set,

$$(\epsilon_i,\epsilon_j) = \delta_{ij} \qquad (i,j = 1,2,\ldots) \tag{2-38}$$

Moreover, since $\{\epsilon_i(x)\}$ is a complete set, then given any $\mathcal{H}$-vector $\psi(x)$, we can find a set of scalars $\{c_i\}$ such that

$$\psi(x) = \sum_{i=1}^{\infty} c_i \epsilon_i(x) \tag{2-39a}$$

Indeed, using Eqs. (2-34) and (2-38), we see that

$$(\epsilon_j, \psi) = \left(\epsilon_j, \sum_{i=1}^{\infty} c_i \epsilon_i\right) = \sum_{i=1}^{\infty} c_i(\epsilon_j, \epsilon_i) = \sum_{i=1}^{\infty} c_i \delta_{ij} = c_j \tag{2-39b}$$

That is, the *expansion coefficients* or *components* c_i of $\psi(x)$ in the orthonormal basis $\{\epsilon_i(x)\}$ are just the scalars (ϵ_i, ψ). Therefore, we can write Eq. (2-39a) as

$$\psi(x) = \sum_{i=1}^{\infty} (\epsilon_i, \psi) \epsilon_i(x) \tag{2-39c}$$

an equation which may be regarded as an identity for all $\psi(x)$ in $\mathcal{H}$, and all orthonormal basis sets $\{\epsilon_i(x)\}$.

Exercise 13. If two $\mathcal{H}$-vectors $\psi(x)$ and $\phi(x)$ have components $\{c_i\}$ and $\{d_i\}$ relative to a given basis $\{\epsilon_i(x)\}$, prove that

$$(\psi, \phi) = \sum_{i=1}^{\infty} c_i^* d_i \tag{2-40a}$$

and in particular that

$$(\psi, \psi) = \sum_{i=1}^{\infty} |c_i|^2 \tag{2-40b}$$

Compare these results with those of Exercise 9.

The properties of the vector space $\mathcal{H}$ are in many respects deeper and subtler than the foregoing development would seem to indicate. However, the depth and rigor of our presentation here will be sufficient for the purposes of this book. The main results of this section are reviewed and summarized in the accompanying table. In the remainder of this book, we shall be concerned only with vectors in $\mathcal{H}$, and not vectors in $\mathcal{E}_3$. However, the correspondences which we have traced between the two vector spaces will often allow us to "visualize," by analogy with $\mathcal{E}_3$, just what it is that we are doing in $\mathcal{H}$. This will help us to keep our feet on the ground, so to speak, as we proceed through the rather abstract theory of quantum mechanics.

Correspondences Between Vectors in $\mathcal{E}_3$ and Vectors in $\mathcal{H}$

Item	$\mathcal{E}_3$	$\mathcal{H}$
Vector	Directed line segment, v	Complex function, $\psi(x)$
Scalar	Real number, r	Complex number, c
Linear combination	$\mathbf{v} = r_1\mathbf{v}_1 + r_2\mathbf{v}_2$	$\psi(x) = c_1\psi_1(x) + c_2\psi_2(x)$
Inner product[1]	$\mathbf{v}_1 \cdot \mathbf{v}_2 \equiv \lvert\mathbf{v}_1\rvert\,\lvert\mathbf{v}_2\rvert \cos\theta_{12}$	$(\psi_1,\psi_2) \equiv \int_{-\infty}^{\infty} \psi_1^*(x)\psi_2(x)dx$
Norm[2]	$\mathbf{v}\cdot\mathbf{v} = \lvert\mathbf{v}\rvert^2$	$(\psi,\psi) = \int_{-\infty}^{\infty} \lvert\psi(x)\rvert^2 dx$
Properties of the inner product	$\mathbf{v}_2\cdot\mathbf{v}_1 = \mathbf{v}_1\cdot\mathbf{v}_2$ $r_1\mathbf{v}_1 \cdot r_2\mathbf{v}_2 = r_1r_2\mathbf{v}_1\cdot\mathbf{v}_2$ $(\mathbf{v}_1+\mathbf{v}_2)\cdot(\mathbf{v}_3+\mathbf{v}_4)$ $= \mathbf{v}_1\cdot\mathbf{v}_3 + \mathbf{v}_1\cdot\mathbf{v}_4 + \mathbf{v}_2\cdot\mathbf{v}_3 + \mathbf{v}_2\cdot\mathbf{v}_4$ $\lvert\mathbf{v}_1\cdot\mathbf{v}_2\rvert \leqq \sqrt{\mathbf{v}_1\cdot\mathbf{v}_1}\sqrt{\mathbf{v}_2\cdot\mathbf{v}_2}$	$(\psi_2,\psi_1) = (\psi_1,\psi_2)^*$ $(c_1\psi_1,c_2\psi_2) = c_1^*c_2(\psi_1,\psi_2)$ $(\psi_1+\psi_2,\psi_3+\psi_4)$ $= (\psi_1,\psi_3) + (\psi_1,\psi_4) + (\psi_2,\psi_3) + (\psi_2,\psi_4)$ $\lvert(\psi_1,\psi_2)\rvert \leqq \sqrt{(\psi_1,\psi_1)}\sqrt{(\psi_2,\psi_2)}$
Orthogonal vectors	$\mathbf{v}_1\cdot\mathbf{v}_2 = 0$	$(\psi_1,\psi_2) = 0$
Orthonormal basis set[3]	$\{\mathbf{e}_i\}$, with $\mathbf{e}_i\cdot\mathbf{e}_j = \delta_{ij}$ and also, for any v in $\mathcal{E}_3$, $\mathbf{v} = \sum_{i=1}^{3}(\mathbf{e}_i\cdot\mathbf{v})\mathbf{e}_i$	$\{\epsilon_i(x)\}$, with $(\epsilon_i,\epsilon_j) = \delta_{ij}$ and also, for any $\psi(x)$ in $\mathcal{H}$ $\psi(x) = \sum_{i=1}^{\infty}(\epsilon_i,\psi)\epsilon_i(x)$

[1] The inner product of two vectors is a *scalar*.
[2] The norm of a vector is a *nonnegative real number*.
[3] The scalars $\{\mathbf{e}_i\cdot\mathbf{v}\}$ and $\{(\epsilon_i,\psi)\}$ are called the *expansion coefficients* or *components* of the vectors v and $\psi(x)$ relative to the respective bases.

2-4 HILBERT SPACE OPERATORS

We recall from elementary calculus that a "function" f is, by definition, a "rule" which associates with each number x another number $y = f(x)$. This concept of a function can be extended to apply to vectors as well as numbers; however, it is then customary to use the term "operator" instead of "function." Thus, $\hat{O}$ is said to be an *operator* in the Hilbert space if and only if $\hat{O}$ specifies some rule of correspondence which associates with each vector $\psi(x)$ in $\mathcal{H}$ another vector $\phi(x)$. We write this as

$$\phi(x) = \hat{O}\psi(x) \tag{2-41}$$

and we speak of $\hat{O}$ as "operating on the vector $\psi(x)$, transforming it into the vector $\phi(x)$."

In the preceding section, we discussed the way in which a *vector* in $\mathcal{H}$ can be (i) multiplied by a scalar, (ii) added to another vector, and (iii) multiplied by another vector. Analogous operations can be defined for *operators* as well. Thus (i) the operator $c\hat{O}$ transforms a given vector $\psi(x)$ into the vector $c(\hat{O}\psi(x))$, (ii) the operator $\hat{O}_1 + \hat{O}_2$ transforms a given vector $\psi(x)$ into the vector $\hat{O}_1\psi(x) + \hat{O}_2\psi(x)$, and (iii) the operator $\hat{O}_1\hat{O}_2$ transforms a given vector $\psi(x)$ into the vector $\hat{O}_1(\hat{O}_2\psi(x))$. Thus, the product of c times $\hat{O}$, the sum of $\hat{O}_1$ and $\hat{O}_2$, and the product of $\hat{O}_1$ times $\hat{O}_2$, are *by definition* such that the following equations are valid for all vectors $\psi(x)$ in $\mathcal{H}$:

$$(c\hat{O})\psi(x) = c(\hat{O}\psi(x)) \tag{2-42a}$$

$$(\hat{O}_1 + \hat{O}_2)\psi(x) = \hat{O}_1\psi(x) + \hat{O}_2\psi(x) \tag{2-42b}$$

$$(\hat{O}_1\hat{O}_2)\psi(x) = \hat{O}_1(\hat{O}_2\psi(x)) \tag{2-42c}$$

In connection with Eq. (2-42c), it must be emphasized that it is *not* necessarily true that $\hat{O}_1\hat{O}_2 = \hat{O}_2\hat{O}_1$; in other words, it is not generally true that, for every $\mathcal{H}$-vector $\psi(x)$, $\hat{O}_1$ acting on $\hat{O}_2\psi(x)$ produces the same vector as does $\hat{O}_2$ acting on $\hat{O}_1\psi(x)$. *If*, however, the equality *does* hold for *all* vectors $\psi(x)$, then we say that $\hat{O}_1$ and $\hat{O}_2$ *commute:*

$$\hat{O}_1 \text{ and } \hat{O}_2 \textit{ commute} \text{ if and only if } \hat{O}_1\hat{O}_2\psi(x) = \hat{O}_2\hat{O}_1\psi(x) \quad \text{for all vectors } \psi(x) \text{ in } \mathcal{H} \tag{2-43}$$

Exercise 14. Let $\hat{O}_1 = x$ [i.e., $\hat{O}_1\psi(x) = x\psi(x)$], and let $\hat{O}_2 = d/dx$ [i.e., $\hat{O}_2\psi(x) = d\psi/dx$]. Show that $\hat{O}_1$ and $\hat{O}_2$ do *not* commute.

In quantum mechanics virtually all operators of interest possess a property called "linearity." By definition, an operator $\hat{O}$ is said to be a *linear* operator if and only if, for any $\mathcal{H}$-vectors $\psi_1(x)$ and

$\psi_2(x)$ and any complex numbers c_1 and c_2,

$$\hat{O}(c_1\psi_1(x) + c_2\psi_2(x)) = c_1\hat{O}\psi_1(x) + c_2\hat{O}\psi_2(x) \qquad (2\text{-}44)$$

Exercise 15. Show that the operator "d/dx" *is* a linear operator, but that the operator "log" is *not* a linear operator.

It is easy to show from Eqs. (2-42) that if $\hat{O}_1$ and $\hat{O}_2$ are linear operators, then the operators $(c_1\hat{O}_1 + c_2\hat{O}_2)$ and $\hat{O}_1\hat{O}_2$ are also linear.

In addition to linearity, another property which many operators in quantum mechanics possess is the property of "hermiticity." An operator $\hat{O}$ is said to be an *Hermitian* operator if and only if, for *any* two $\mathcal{H}$-vectors $\psi_1(x)$ and $\psi_2(x)$,

$$(\psi_1, \hat{O}\psi_2) = (\hat{O}\psi_1, \psi_2) \qquad (2\text{-}45)$$

As an example, let us see if the simple operator $\hat{O} = c$ is an Hermitian operator. If $\psi_1(x)$ and $\psi_2(x)$ are any two $\mathcal{H}$-vectors, then, using Eq. (2-34b), we have

$$(\psi_1, c\psi_2) = c(\psi_1, \psi_2) = (c^*\psi_1, \psi_2)$$

Thus, according to Eq. (2-45), the operator $\hat{O} = c$ is Hermitian if and only if $c = c^*$—i.e., if and only if c is real.

Exercise 16. If $\hat{O}_1$ and $\hat{O}_2$ are Hermitian operators, prove that
(a) $c_1\hat{O}_1 + c_2\hat{O}_2$ is Hermitian if c_1 and c_2 are real.
(b) $\hat{O}_1\hat{O}_2$ is Hermitian if $\hat{O}_1$ and $\hat{O}_2$ commute.

We turn now to discuss one final aspect of operators which will prove to be very essential to the mathematical formulation of quantum mechanics. We make the following definition: If the effect of a given operator $\hat{O}$ on some *particular* $\mathcal{H}$-vector $\psi(x)$ is to simply multiply that vector by an $\mathcal{H}$-scalar c,

$$\hat{O}\psi(x) = c\psi(x) \qquad (2\text{-}46)$$

then we say that the vector $\psi(x)$ is an *eigenvector* (or eigenfunction) of $\hat{O}$, and c is the corresponding *eigenvalue.*†

Exercise 17.
(a) Show that the function e^{ax} (where a is real) is an eigenfunction of the operator "d/dx." What is the corresponding eigenvalue?
(b) Show that the function x^n (where $n \geqq 1$) is an eigenfunc-

† The prefix "eigen" is a German word. When we call c an *eigenvalue* of $\hat{O}$, we mean literally that c is a value which is "characteristic of" or "peculiar to" the operator $\hat{O}$.

tion of the operator "$x \cdot d/dx$." What is the corresponding eigenvalue?

(c) Of what operator (excluding $\hat{O} = c$) is the function $\cos ax$ an eigenfunction?

We can now establish two important results concerning the eigenvectors and eigenvalues of *Hermitian* operators:

(i) The eigenvalues of an Hermitian operator are pure real. To see this, suppose $\hat{O}$ is an Hermitian operator with eigenvector $\psi(x)$ and eigenvalue c. By Eqs. (2-46) and (2-34b), we can write

$$(\psi,\hat{O}\psi) = (\psi,c\psi) = c(\psi,\psi)$$

and

$$(\hat{O}\psi,\psi) = (c\psi,\psi) = c^*(\psi,\psi)$$

But, since $\hat{O}$ is Hermitian, then these two quantities must be equal:

$$c(\psi,\psi) = c^*(\psi,\psi)$$

Excluding the trivial case in which $\psi(x)$ is the null vector, we have $(\psi,\psi) > 0$, so we may conclude that $c = c^*$—i.e., c is pure real.

(ii) The eigenvectors corresponding to two unequal eigenvalues of an Hermitian operator are orthogonal to each other. The proof of this statement is the subject of the following exercise.

Exercise 18. Let $\hat{O}$ be an Hermitian operator with eigenfunctions $\psi_1(x)$ and $\psi_2(x)$, and let the corresponding eigenvalues c_1 and c_2 be unequal. Prove that $\psi_1(x)$ and $\psi_2(x)$ are orthogonal. [*Hint:* Consider the two quantities $(\psi_1,\hat{O}\psi_2)$ and $(\hat{O}\psi_1,\psi_2)$, and use the fact just established that c_1 and c_2 must be pure real.]

We shall now prove a theorem that is almost, but not quite, the converse of the preceding two theorems, (i) and (ii). Suppose $\hat{A}$ is a linear operator which possesses a *complete*, *orthonormal set of eigenvectors* $\{\alpha_n(x)\}$ and a corresponding *set of real eigenvalues* $\{a_n\}$:

$$\left.\begin{aligned} \hat{A}\alpha_n(x) &= a_n\alpha_n(x), \quad a_n \text{ real} \\ (\alpha_m,\alpha_n) &= \delta_{mn}, \quad \text{all } m,n \end{aligned}\right\} \tag{2-47}$$

We shall prove that these conditions imply that the operator $\hat{A}$ is *Hermitian*. We shall do this by showing that, for any two $\mathcal{H}$-vectors $\psi(x)$ and $\phi(x)$, it is true that $(\psi,\hat{A}\phi) = (\hat{A}\psi,\phi)$.

Define $c_n = (\alpha_n,\psi)$ and $e_n = (\alpha_n,\phi)$, and expand $\psi(x)$ and $\phi(x)$

in the orthonormal basis $\{\alpha_n(x)\}$ according to Eqs. (2-39):

$$\psi(x) = \sum_n c_n \alpha_n(x) \qquad \phi(x) = \sum_n e_n \alpha_n(x)$$

Using the linearity of Â, the properties of the inner product in Eqs. (2-34), and the orthonormality of the set $\{\alpha_n(x)\}$, we have

$$(\hat{A}\psi, \phi) = (\hat{A}\sum_n c_n \alpha_n, \sum_m e_m \alpha_m)$$

$$= \left(\sum_n c_n \mathrm{A}\alpha_n, \sum_m e_m \alpha_m\right)$$

$$= \left(\sum_n c_n a_n \alpha_n, \sum_m e_m \alpha_m\right)$$

$$= \sum_{m,n} c_n^* a_n^* e_m (\alpha_n, \alpha_m)$$

$$= \sum_{m,n} c_n^* a_n^* e_m \delta_{nm}$$

so

$$(\hat{A}\psi, \phi) = \sum_n c_n^* a_n^* e_n$$

In an exactly analogous way, we find that

$$(\psi, \hat{A}\phi) = \sum_n c_n^* e_n a_n$$

Exercise 19. Carry out the steps leading to this last equation.

Since the eigenvalues $\{a_n\}$ were given to be real, $a_n^* = a_n$, it follows at once that $(\hat{A}\psi, \phi) = (\psi, \hat{A}\phi)$; therefore, Â is Hermitian.

The foregoing theorem and its proof should be studied in detail. Not only is its content important, but the steps in its proof illustrate well the sort of mathematical manipulations which will be required in our development of quantum mechanics.

Exercise 20. Suppose a linear operator Â has a complete, orthonormal set of eigenvectors $\{\alpha_n(x)\}$ and a corresponding set of eigenvalues $\{a_n\}$. Show that a knowledge of the elements of the two sets $\{\alpha_n(x)\}$ and $\{a_n\}$ is sufficient to completely specify the operator. [*Hint:* Show that, for any *given* $\mathcal{H}$-vector $\psi(x)$, the vector $\hat{A}\psi(x)$ is completely defined through the quantities $\{\alpha_n(x)\}$ and $\{a_n\}$.]

CHAPTER

3

A Brief Review of Classical Mechanics

An important prerequisite for a meaningful understanding of quantum mechanics is a clear appreciation of the fundamental principles of classical mechanics. It is assumed that the reader is already familiar with the more elementary ideas and attitudes of classical mechanics. In this chapter we shall simply try to *organize*, and occasionally to *expand*, these ideas and attitudes, in a way that will best enable us to see later the basic similarities and differences between classical and quantum mechanics. At the same time, the level and approach of our development of the classical theory in this chapter should give the reader a rough indication of the level and approach of our development of the quantum theory in the next chapter.

3-1 A MECHANICAL SYSTEM

Our objects of study will be "mechanical systems" which, for the sake of simplicity, have only one degree of freedom. For concreteness, we shall take as our system a single particle of constant mass m which is constrained to move along the x-axis in a "conservative" force field, $F(x)$. In order to avoid the complications of the theory of relativity, we restrict our discussion to those cases in which the velocity of the particle,

$$v \equiv \frac{dx}{dt} \tag{3-1}$$

is always much less than the velocity of light.

The *force function* $F(x)$ gives the force exerted on the particle by its environment at each point x; thus the force function may be said to describe the mechanical interaction of the particle with its environment. This interaction may also be described by the *potential function* $V(x)$, which *by definition* is that function whose negative derivative with respect to x is the force function:

$$F(x) \equiv -\frac{d}{dx} V(x) \tag{3-2}$$

Although it is possible to think of forces which cannot be derived from any function $V(x)$ according to Eq. (3-2) (e.g., a *frictional* force, which depends not on x, but rather on the direction of motion), our earlier stipulation that the force field be "conservative" means precisely that $V(x)$, as defined by Eq. (3-2), *does* exist. Consequently, for the systems of interest to us it makes no difference whatsoever whether we describe the interaction between the particle and its environment by specifying $F(x)$ or $V(x)$, since if one of these functions is known the other one may be found through Eq. (3-2).

The physical significance of the potential function may be understood as follows: Consider the graph of $V(x)$ in a small neighborhood of some point x_0. If the graph is sloping *upward* in this neighborhood, then $dV/dx|_{x_0} > 0$, and so Eq. (3-2) implies that the force $F(x_0)$ is in the *negative* x-direction; on the other hand, if the graph is sloping *downward*, then $dV/dx|_{x_0} < 0$, and Eq. (3-2) implies that the force $F(x_0)$ is in the *positive* x-direction. Thus, the force $F(x)$ always *tries* to move the particle in that direction which would result in a *decrease* in $V(x)$; moreover, Eq. (3-2) says that the strength or *magnitude* of this force at a given point is numerically equal to the *rate of decrease* of $V(x)$ at that point. We may therefore think of the graph of $V(x)$ as being a sort of "hilly terrain" upon which the particle rolls under the influence of some pseudogravitational force.

Exercise 21.

(a) Show that two potential functions which differ by only a constant (i.e., $V_2(x) = V_1(x) + C$) give rise to identical force functions, and hence are "physically equivalent" potentials.

(b) For the force field $F(x) = -kx$, what is the potential field? For the potential field $V(x) = k/x$, what is the force field?

(c) If the point x_0 is a *local minimum* of $V(x)$, show that x_0 is a point of *stable equilibrium*; i.e., show that the particle feels *no* force at the point x_0, while at any point slightly above or below x_0 the particle feels a force acting *toward* x_0. In a similar way, show that if x_0 is a *local maximum* of $V(x)$, then x_0 is a point of *unstable equilibrium*.

The basic program of mechanics, both classical and quantum, is essentially twofold: First, we have to decide how we shall specify the *instantaneous state* of a given mechanical system, and then we must discover how this state changes or *evolves with time*. In this chapter we shall review how classical mechanics accomplishes these ends, and in the next chapter we shall consider the approach taken by quantum mechanics.

3-2 THE CLASSICAL STATE

In classical mechanics the instantaneous state of a mechanical system is described in terms of the values of certain "observable variables" of the system. For the simple system of a particle of mass m constrained to move along the x-axis, the observable variables used to define the state are normally the *position* x and the *momentum* $p \equiv mv$ of the particle. In other words, the state of the system at time t is specified by the pair of values $[x(t),p(t)]$.

This classical definition of the state of a mechanical system tacitly assumes that:

(i) the position and momentum variables both have precise, well-defined values at each instant of time; and

(ii) it is always possible, at least in principle, to measure these values without significantly disturbing the system.

These assumptions might seem so natural and innocuous as to hardly merit mentioning. However, we shall find in the next chapter that quantum mechanics, in its most widely accepted formulation, actually *denies* the general validity of *both* these assumptions! This comes about as a consequence of the radically different viewpoint taken by quantum mechanics with regard to the concepts of "state" and "observable variables," and to the role played by the measurement process. An elucidation of these important points will occupy much of our attention in the following chapter.

3-3 TIME EVOLUTION OF THE CLASSICAL STATE

3-3a The Newtonian Formulation

Having defined the instantaneous state of our classical, one-particle system, we must now address ourselves to the problem of discovering how the state changes with time. The assumption of course is that the state variables x and p stand in a definite functional relationship to the time variable t, and our object is to determine the precise *forms* of the functions $x(t)$ and $p(t)$. One way to achieve these ends is to *postulate* Newton's second law,

$$\text{Force} = \text{mass} \times \text{acceleration}$$

which we shall write here in the form

$$\frac{d^2x}{dt^2} = \frac{F(x)}{m} \tag{3-3a}$$

In words, this law says that the desired function $x(t)$ is that function whose second time derivative is equal to the force function divided by the particle mass. Evidently, then, to find $x(t)$ we must integrate Eq. (3-3a) twice with respect to t; once $x(t)$ is found, we then obtain the function $p(t)$ by differentiating $x(t)$ and multiplying by m:

$$p = m\frac{dx}{dt} \tag{3-3b}$$

To be sure, the twofold integration of Newton's second law may be very difficult, if not impossible, to perform analytically for certain force functions $F(x)$. However, this is more a problem of applied mathematics rather than physics (although it sometimes happens that a "feeling" for a given physical problem will suggest a fruitful way of carrying out the requisite integrations). For our purposes, though, we merely content ourselves with the thought that Eq. (3-3a) *in principle* determines the function $x(t)$ for any given force field $F(x)$, regardless of how difficult it may be to explicitly *solve* the differential equation.

In twice integrating Eq. (3-3a), we will generate two constants of integration. The values of these two constants may be uniquely fixed by specifying the values of x and dx/dt at some "initial time" $t = 0$. Equivalently, since dx/dt and p are related by Eq. (3-3b), we may specify the initial value of p instead of dx/dt.

In conclusion we see that, with either $F(x)$ or $V(x)$ given, and with the "initial state" $[x(0), p(0)]$ specified, then Eqs. (3-3a) and (3-3b) enable us to determine unambiguously the state of the system $[x(t),p(t)]$ at any time $t > 0$.†

†If the reader wondered why, in the previous section, we chose to specify the state of our system by the *pair* of variables $[x(t),p(t)]$, rather than by $x(t)$ alone, the reason should now be apparent: since the time evolution equation for $x(t)$, Eq. (3-3a), contains a *second*-order time derivative, then a specification of $x(0)$ alone would *not* suffice to unambiguously determine $x(t)$ for all $t > 0$. Generally speaking, the observable variables chosen to specify the "classical state" of a system must be such that their initial values collectively determine their subsequent values.

As a familiar example, for the simple force field $F(x) = k$, or $V(x) = -kx$, Newton's second law reads

$$\frac{d^2x}{dt^2} = \frac{k}{m}$$

The two t-integrations yield successively

$$\frac{dx}{dt} = \frac{k}{m} t + C_1, \qquad x = \frac{1}{2}\frac{k}{m} t^2 + C_1 t + C_2$$

Hence,

$$x(t) = \frac{1}{2}\frac{k}{m} t^2 + C_1 t + C_2, \qquad p(t) = m\frac{dx}{dt} = kt + mC_1$$

The requirements that $x(0) = x_0$ and $p(0) = p_0$ imply that $C_2 = x_0$ and $mC_1 = p_0$. Consequently, if $[x_0, p_0]$ is the state at time 0, then the state at time t is evidently

$$\left[\frac{1}{2}\frac{k}{m} t^2 + \frac{p_0}{m} t + x_0, \qquad kt + p_0\right]$$

3-3b Energy

One very important consequence of Newton's second law is the introduction of the concept of energy. This concept arises naturally through a consideration of the quantity

$$W_{12} \equiv \int_{x_1}^{x_2} F(x)dx \tag{3-4}$$

which is called "the work done on the particle by the force $F(x)$ during the motion from x_1 to x_2." Of course this definition, like any definition, tells us nothing new; however, let us use the two expressions for $F(x)$ given in Eqs. (3-2) and (3-3a) to calculate two different expressions for W_{12}, and then see what we can learn by equating the results. First, from the definition of the potential function in Eq. (3-2), we have

$$W_{12} = \int_{x_1}^{x_2} \left(-\frac{dV}{dx}\right)dx = -\int_{x_1}^{x_2} dV = -V(x)\Big|_{x_1}^{x_2}$$

so

$$W_{12} = -[V(x_2) - V(x_1)] \tag{3-5a}$$

Second, from Newton's second law in Eq. (3-3a), making use of Eq. (3-1) and the chain rule for derivatives, we have

$$W_{12} = \int_{x_1}^{x_2} \left(m \frac{d^2 x}{dt^2}\right) dx = \int_{x_1}^{x_2} \left(m \frac{dv}{dt}\right) dx = \int_{x_1}^{x_2} \left(m \frac{dv}{dx}\frac{dx}{dt}\right) dx$$

$$= \int_{x_1}^{x_2} \left(mv \frac{dv}{dx}\right) dx = \int_{v_1}^{v_2} mv dv = \frac{1}{2} mv^2 \Big|_{v_1}^{v_2}$$

so

$$W_{12} = \frac{1}{2} mv_2^2 - \frac{1}{2} mv_1^2 \tag{3-5b}$$

where v_1 and v_2 are the velocities of the particle at x_1 and x_2 respectively. Equating the two expressions for W_{12} in Eqs. (3-5a) and (3-5b), we find that

$$\frac{1}{2} mv_1^2 + V(x_1) = \frac{1}{2} mv_2^2 + V(x_2) \tag{3-5c}$$

But since the points x_1 and x_2 were completely arbitrary, then we may conclude that the quantity

$$E \equiv \frac{1}{2} mv^2 + V(x) \tag{3-6a}$$

maintains a constant value throughout the motion of the particle. E is called the *energy* of the system. It is seen to be composed of two parts, one part associated with the *motion*, $mv^2/2$, and the other part associated with the *position*, $V(x)$; evidently, the particle moves in such a way that any decrease in one of these terms is always exactly compensated by an increase in the other term.

To demonstrate the usefulness of the energy concept in studying the dynamics of a mechanical system, we show in Fig. 2 a plot of $V(x)$ versus x for a particle in some hypothetical force field. If the total energy is E, then the particle is constrained to move only in those regions in which $V(x) < E$, since the quantity $mv^2/2$ in Eq. (3-6a) cannot go negative. In such a region, the distance from the curve $V(x)$ up to the line $V = E$ is evidently just the difference between the *total* energy E and the *potential* energy $V(x)$—i.e., this distance is just the *kinetic* energy $mv^2/2$. In Fig. 2 we have illustrated how the energy of the particle at a given point is divided between the kinetic and potential terms. The points $x = a$ and $x = b$ satisfy $V(a) = V(b) = E$, and define the *boundaries* of the motion; at these points the velocity of the particle must obviously vanish, and

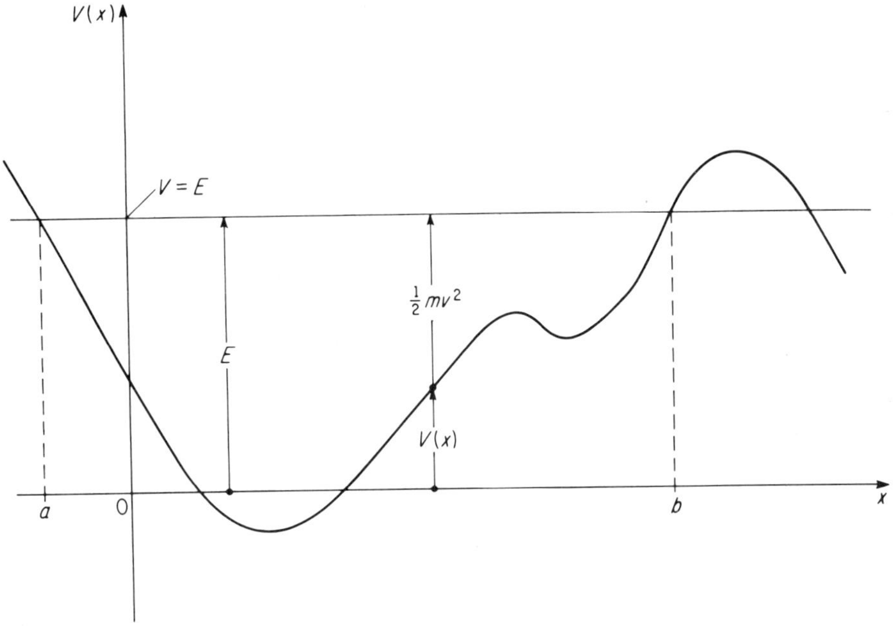

Fig. 2. A plot of the potential function $V(x)$ versus x for a particle in some hypothetical force field. The distance from the curve $V(x)$ up to the horizontal line $V = E$ represents the kinetic energy, $\frac{1}{2}mv^2$. Since this quantity cannot become negative, the motion of the particle is confined to the interval $a < x < b$, where $V(x) < E$. The distance from the base line up to the curve represents the potential energy $V(x)$, which can be either positive or negative.

the particle will evidently reverse its direction of travel. Thus the energy plot in Fig. 2 indeed presents a very clear and concise picture of the motion of the particle: the distance $(E - V(x))$ provides a measure of the *speed* of the particle at the point x, while the slope of the potential function curve at the point x provides a measure of the direction and magnitude of the *force* on the particle, or the *acceleration* of the particle, at this point.

Since $p = mv$, then we can also write E as

$$E = \frac{p^2}{2m} + V(x) \tag{3-6b}$$

The fact that *this particular function* of the state variables x and p remains *constant* as these two variables evolve with time is clearly a non-trivial consequence of Newton's second law, and is why the concept of energy occupies such an important place in the structure of classical mechanics.

Exercise 22. Calculate dE/dt directly from its definition in Eq. (3-6a) and show that, as a consequence of Newton's second law and the definition of the potential function, $dE/dt = 0$. [*Hint*: Since $V(x)$ depends on t only implicitly through x, use the chain rule to calculate dV/dt.]

3-3c The Hamiltonian Formulation

Equations (3-3a) and (3-3b) comprise what is called the "Newtonian formulation" of classical mechanics; it is this formulation that is usually presented in introductory physics courses. It so happens that there are several other ways to formulate the time evolution of the state variables $x(t)$ and $p(t)$; of course, all these ways are entirely equivalent in physical content to the Newtonian formulation, but certain of them offer special advantages in various situations. The so-called "Hamiltonian formulation" is especially relevant to our purposes because it helps us to see a little more clearly the relation between classical and quantum mechanics.

From a strictly classical point of view, the Hamiltonian formulation is nice because it places the state variables x and p on a formally equivalent footing. We note that, by contrast, the Newtonian formulation treats x as the primary variable [see Eq. (3-3a)], while p seems to be merely an auxiliary variable which is derived from x [see Eq. (3-3b)]. For the particular case of a particle on the x-axis in a potential field $V(x)$, the transition from the Newtonian to the Hamiltonian formulation can be easily made: Using Eqs. (3-1) and (3-2), we can write Eqs. (3-3b) and (3-3a) respectively as

$$\frac{dx}{dt} = \frac{p}{m} \tag{3-7a}$$

and

$$\frac{dp}{dt} = -\frac{dV}{dx} \tag{3-7b}$$

We now *define* the *Hamiltonian function* $H(x,p)$ to be the total energy of the system expressed as a function of the state variables x and p. Thus, using Eq. (3-6b), we define

$$H(x,p) \equiv \frac{p^2}{2m} + V(x) \tag{3-8}$$

We now observe that the partial derivatives† of $H(x,p)$ with respect to its two arguments are given by

$$\frac{\partial H}{\partial p} = \frac{p}{m} \quad \text{and} \quad \frac{\partial H}{\partial x} = \frac{dV}{dx} \tag{3-9}$$

Exercise 23.

(a) Show that Eqs. (3-7) are equivalent to Newton's equations, (3-3).

(b) Show that Eqs. (3-9) follow from our definition of the Hamiltonian function.

It is now a simple matter to combine Eqs. (3-9) with Eqs. (3-7) to obtain

$$\frac{dx}{dt} = \frac{\partial}{\partial p} H(x,p) \tag{3-10a}$$

$$\frac{dp}{dt} = -\frac{\partial}{\partial x} H(x,p) \tag{3-10b}$$

These two equations are called "Hamilton's equations of motion," and they constitute the Hamiltonian formulation of classical mechanics in the same way that Eqs. (3-3a) and (3-3b) constitute the Newtonian formulation. It should be clear from our derivation of Hamilton's equations that they contain no more nor no less information than Newton's equations. However, it is equally clear that the state variables x and p *formally* play more symmetric roles in Eqs. (3-10) than in Eqs. (3-3). From a strictly mathematical point of view, what we have done is to replace a single, second-order differential equation in the one variable x [Eq. (3-3a)] with two coupled, first-order differential equations in the two variables x and p [Eqs. (3-10)]. It should also be noted that, in the Hamiltonian formulation, the mechanical interaction of the particle with its environment is now formally described by the Hamiltonian function, rather than by the force function or the potential function. Of course, it is clear from the definitions in Eqs. (3-2) and (3-8) that a knowledge of any one of the functions $H(x,p)$, $V(x)$, or $F(x)$, implies a knowledge of the other two.

To find the functions $x(t)$ and $p(t)$ via the Hamiltonian formu-

†If f is a function of two variables u and v, then $\partial f/\partial u$, the "partial derivative of f with respect to u," is defined to be the u-derivative taken with v treated as a constant. Thus if $f(u,v) = u^2 v^3$, then we have $\partial f/\partial u = 2uv^3$, and $\partial f/\partial v = 3v^2 u^2$.

lation, we must evidently integrate Eqs. (3-10a) and (3-10b) once each with respect to t. Again, the actual integrations might be quite difficult to *perform*—especially since the equations are *coupled* (i.e., x and p both appear in both equations); however, we are not concerned here with these "practical details." Of more interest to us is the fact that the two integrations will yield two integration constants, and, as in the Newtonian formulation, these two constants can be fixed by specifying the values of x and p at some "initial time" $t = 0$. Thus, with $H(x,p)$ given, and with the initial state $[x(0),p(0)]$ specified, then Eqs. (3-10a) and (3-10b) enable us to determine unambiguously the state of the system $[x(t),p(t)]$ at any time $t > 0$.

Exercise 24. We have derived Hamilton's equations from Newton's equations. Prove now that Newton's equations can be derived from Hamilton's equations. That is, show that Eqs. (3-3) follow from Eqs. (3-10) when account is taken of the definitions of $H(x,p)$ and $V(x)$.

3-3d "Determinism" in Classical Mechanics

Classical mechanics has sometimes been said by philosophers to imply a "deterministic" universe. By this it is meant that, given the initial state of the universe (i.e., given the exact positions and momenta of all the particles in the universe at some time $t = 0$), and given also the functional forms of all the forces acting on and among these particles, then the subsequent "history" of the universe is in principle completely *determined* through the dynamical equations of Newton (or Hamilton). We shall see in the next chapter that the tenets of quantum mechanics will force us to dramatically revise (but not to completely discard) this simple deterministic picture of the physical universe.

CHAPTER

4

The Theory of Quantum Mechanics

The resumé of classical mechanics given in the last chapter was presented from a very elementary point of view: We did not attempt to critically analyze the basic concepts of length, time, force and mass; we confined ourselves to a system so simple that it precluded any discussion of such important topics as angular momentum and multiparticle phenomena; we did not present the Hamiltonian formulation in its full generality, nor even mention any other formulations; and we did not try to use classical mechanics to solve various specific "problems." In truth, our sole aim was to present, in as simple and uncomplicated a way as possible, only the barest essentials of the classical theory. And in spite of (or maybe because of) all our omissions, we did gain a concise and fairly accurate perspective of the general aims, assumptions and methodology of classical mechanics. In the present chapter we shall try to carry out an analogous program with respect to quantum mechanics.

Unfortunately, quantum mechanics is inherently abstract, and is not as easy to grasp and understand as classical mechanics. It might seem odd that, in their attempt to formulate a more truthful picture of physical phenomena, physicists came up with a theory which is so highly abstract that it seems quite remote from physical reality. Yet we must bear in mind that our personal notions of physical reality are derived from our lifelong contact with *macroscopic* phenomena, and we have no license at all to extrapolate our "macroscopic intuition" to the *microscopic* level. Indeed, the physicist has found, through carefully performed experiments, that microscopic reality is characterized by phenomena such as the wave-particle duality which seem quite unintelligible in terms of "common sense" reasoning. We should not be too surprised, then, to find that the logical system which purports to account for these seeming anomalies will itself seem foreign to our deeply ingrained, classical point of view. As a consequence, the reader will be obliged to simply accept the abstractness of the quantum theory, because at the present time there does not seem to be any other way to get hold of it. Whether this is a consequence of "the way things are," or whether this indicates that our present understanding of quantum mechanics is in some way deficient, is a question which many physicists are still pondering.

We are going to present the theory of quantum mechanics by laying down a number of *postulates* (six, to be precise), from which we shall deduce various consequences. These postulates will undoubtedly seem strange to the reader, and completely devoid of any intuitive appeal. However, the reader must keep in mind that, together, these postulates form the simplest and most widely accepted logical basis yet devised for understanding *quantitatively* an enormous range of physical phenomena; indeed, it is this fact, and not any "reasonableness" on the part of the postulates themselves, that gives us cause to accept them. The postulates, together with their derived consequences, will constitute our "picture" of quantum mechanics.

It should be noted that both the number and content of the fundamental postulates of quantum mechanics are to some extent a matter of personal taste. In keeping with the limited aims of this book, the postulates presented here will not be as general and logically economical as they could be, and of course, we shall only attempt to derive a limited selection of their consequences.

We shall develop the theory of quantum mechanics for a nonrelativistic physical system with one degree of freedom, which we represent by the variable x. Although our treatment in Secs. 4-1 through 4-4 will be applicable to *any* kind of nonrelativistic system with one degree of freedom, the reader may find it helpful to keep in mind the specific system discussed in the preceding chapter—i.e., a particle of mass m moving on the x-axis in a potential field $V(x)$. We shall specialize our treatment to this important type of system in Sec. 4-5.

4-1 THE QUANTUM STATE

In Chapter 3 we saw that classical mechanics identifies the *state* of a physical system with the current values of certain *observables* of the system (e.g., the position x and the momentum p). Quantum mechanics, on the other hand, makes a very sharp distinction between states and observables. Concerning the *state* of a system in quantum mechanics, we have the following postulate:

Postulate 1. Every possible physical state of a given system corresponds to some normed Hilbert space vector $\psi(x)$, and conversely, every normed Hilbert space vector $\psi(x)$ corresponds to a possible physical state of the system. This correspondence between physical states and normed vectors in $\mathcal{H}$ is one-to-one, except that two normed $\mathcal{H}$-vectors that differ only by an overall scalar factor of modulus unity correspond to the same physical state. The particular $\mathcal{H}$-vector to which the state of the system corresponds at time t is denoted by $\Psi_t(x)$ and is called the *state vector* of the system; the system is said to "be in the state $\Psi_t(x)$." The state of a system is *completely described* by the state vector in the sense that anything which is in principle knowable about the system at time t can be learned from the function $\Psi_t(x)$.

This postulate makes three assertions: First, it asserts that the possible physical states of a given system stand in a one-to-one correspondence with the normed $\mathcal{H}$-vectors $\psi(x)$ which are defined *up to* an overall scalar factor of modulus unity. In saying that $\psi(x)$ is "normed," we mean simply that its norm, or inner product with itself, equals 1: $(\psi,\psi) = 1$. In saying that $\psi(x)$ is defined up to an overall scalar factor of modulus unity, we mean that if c is any $\mathcal{H}$-scalar (i.e., any complex number) satisfying $|c|^2 = 1$, then the two $\mathcal{H}$-vectors $\psi(x)$ and $\phi(x) = c\psi(x)$ are "physically equivalent" in that they correspond to the same physical state.

Exercise 25. If $\psi(x)$ has unit norm, and c is a complex number satisfying $|c|^2 = 1$, show that $\phi(x) = c\psi(x)$ also has unit norm.

The second assertion of Postulate 1 is that everything that can possibly be known about the state of the system at time t can be obtained from its "state vector" $\Psi_t(x)$. However, the postulate says nothing about *what* things can be known, nor *how* they can be derived from the state vector. These questions will evidently have to be answered by subsequent postulates.

In view of these first two assertions of Postulate 1, we can conclude that the state of a system at time t is completely specified if and only if the state *vector* $\Psi_t(x)$ is given as a definite *function of* x. For comparison, we recall that in classical mechanics the state of a system at time t is completely specified if and only if the state *variables* (e.g., $x(t)$ and $p(t)$) are given as definite *real numbers.* In this connection, the reader should *not* try to identify the x in $\Psi_t(x)$ with the classical position variable $x(t)$; that is, $\Psi_t(x)$ does *not* stand for $\Psi(x(t))$. The x in $\Psi_t(x)$ merely represents the argument of the function Ψ_t; it is a dummy variable, and does not depend in any way upon the time variable t. In the special case where the physical system under consideration is a mass particle, it turns out that there is a very intimate connection between "position" and the argument of the Hilbert space vectors; however, this connection is quite unlike anything the reader might now suppose, and it cannot be explained until after all the postulates of quantum mechanics have been introduced.

The third assertion of Postulate 1 is somewhat indirect. In writing the state vector of the system at time t as $\Psi_t(x)$, it is implied that the state vector is in some sense a function of time. To bring this out more explicitly, we shall sometimes write $\Psi_t(x)$ as $\Psi(x,t)$. It must be emphasized, however, that the functional dependence of Ψ on t is essentially different from its dependence on x: as a Hilbert space vector, $\Psi(x,t)$ is properly a function of x alone, and the parameter t serves merely to label different vectors in $\mathcal{H}$. Thus, $\Psi(x,t_1)$ and $\Psi(x,t_2)$ are to be regarded as two different $\mathcal{H}$-vectors—i.e., two different functions of x—which specify the state of the system at two different times t_1 and t_2. The behavior of $\Psi_t(x) \equiv \Psi(x,t)$ as a function of time—i.e., the "time evolution" of the state vector—will be taken up in Postulate 5 [Sec. 4-4]. Until then we shall be mainly concerned with the state of the system at a single instant t.

Finally, we should mention that the state vector $\Psi_t(x)$ is sometimes referred to as the "state function" or "wave function" of the system.

4-2 OBSERVABLES IN QUANTUM MECHANICS

The quantum mechanical specification of the *state* of a system evidently makes no reference at all to any physical *observables* of the system; this is in marked contrast to the way in which the state of a system is specified in classical mechanics [see Sec. 3-2]. However, Postulate 1 does assert that the state vector in some way contains everything we can possibly know about the system. It therefore seems reasonable to expect that, if at some time t we know the exact functional form of the state vector $\Psi_t(x)$, then we ought to be able to make some fairly definite assertions about the physical observables of the system at that instant; indeed, if we could not do this, then the state vector would be a completely useless mathematical abstraction. Before presenting the postulate which tells us exactly what *can* be said about an observable when the state vector is known, it is first necessary to define more precisely the quantum mechanical concept of an observable. Since for the present we want to keep our discussion as general as possible, we shall refer to observables by script capital letters (e.g., the observable $\mathcal{A}$, the observable $\mathcal{B}$, etc.).

As in classical mechanics, an *observable* $\mathcal{A}$ is simply a dynamical variable that can be measured; e.g., for a mass on the x-axis, the observables are the position, the momentum, and functions of the position and momentum (of which the energy is perhaps the most useful). The *measurement* of an observable $\mathcal{A}$ is some well-defined physical operation which, when performed on the system, yields a single real number called "the value of $\mathcal{A}$." For simplicity, we shall consider all measurements to be "ideal" in the sense that the measured value has an experimental uncertainty of zero.

Now in classical mechanics, no real distinction is made between the *mathematical representation* of an observable and the *values* of the observable; however, in quantum mechanics such a distinction is of fundamental importance. Postulate 2, which we now state, is concerned with (*a*) the mathematical representation of $\mathcal{A}$, and (*b*) the possible values of $\mathcal{A}$.

Postulate 2.

(a) To each physical observable $\mathcal{A}$, there corresponds in the Hilbert space a linear Hermitian operator $\hat{A}$, which possesses a complete, orthonormal set of eigenvectors $\alpha_1(x)$, $\alpha_2(x)$, $\alpha_3(x)$, . . . and a corresponding set of real eigenvalues A_1, A_2, A_3, . . .

$$\hat{A}\alpha_i(x) = A_i\alpha_i(x) \qquad i = 1,2,3 \ldots \tag{4-1}$$

Conversely, to each such operator in the Hilbert space there corresponds some physical observable.

(b) The only possible values which any measurement of $\mathcal{A}$ can yield are the eigenvalues $A_1, A_2, A_3, \ldots$.

Let us begin our discussion of this postulate by reviewing the definitions of the mathematical terms used in part (a). The fact that the $\mathcal{H}$-vectors $\{\alpha_i(x)\}$ are "eigenvectors" of the operator $\hat{A}$ with "eigenvalues" $\{A_i\}$ simply means that Eqs. (4-1) hold true for all i. To say that the eigenvectors form a "complete, orthonormal set" means that, for all i and j,

$$(\alpha_i, \alpha_j) \equiv \int_{-\infty}^{\infty} \alpha_i^*(x)\,\alpha_j(x)\,dx = \delta_{ij} \tag{4-2a}$$

and, moreover, that any $\mathcal{H}$-vector $\phi(x)$ can be written as

$$\phi(x) = \sum_{i=1}^{\infty} (\alpha_i, \phi)\alpha_i(x) \tag{4-2b}$$

In other words, the eigenvector set $\{\alpha_i(x)\}$ is an "orthonormal basis set" in the Hilbert space. For brevity, we shall refer to the set $\{\alpha_i(x)\}$ as the *eigenbasis* of $\hat{A}$. The condition that the eigenvalues $\{A_i\}$ all be real can be conveniently written

$$A_i^* = A_i \tag{4-3}$$

Finally, the fact that $\hat{A}$ is an "Hermitian" operator means that, for any two $\mathcal{H}$-vectors $\phi_1(x)$ and $\phi_2(x)$,

$$(\phi_1, \hat{A}\phi_2) = (\hat{A}\phi_1, \phi_2) \tag{4-4}$$

It should be noted that the Hermiticity of $\hat{A}$ really need not have been postulated, because we proved in Chapter 2 [see Exercise 19] that any linear operator which possesses a complete, orthonormal set of eigenvectors and a corresponding set of real eigenvalues is *necessarily* Hermitian.† Nevertheless, the Hermitian property is such a fundamental property of observable operators that we have allowed this minor redundancy to enter into our statement of Postulate 2.

Postulate 2 says, first of all, that a physical observable $\mathcal{A}$ is "mathematically represented" by a linear operator $\hat{A}$ which possesses a complete, orthonormal set of eigenvectors and a corresponding set of real eigenvalues. We shall call such an operator an *observable operator*. The consequences of mathematically representing observables by operators remain to be seen, but it is reasonable to expect that these consequences will depend strongly on the mathematical rules for manipulating operators. These rules were discussed in Sec. 2-4. Perhaps the most striking difference between these rules and the rules for manipulating ordinary numbers is that, whereas two numbers A and B always *commute* ($AB \equiv BA$), it is *not* true that two operators $\hat{A}$ and $\hat{B}$ necessarily commute [e.g., see Exercise 14]. Thus, in some sense it may be said that observables always commute in classical mechanics, but not in quantum mechanics. We shall see later that the fact that not all pairs of observable operators commute

†A linear operator can be Hermitian *without* possessing a complete, orthonormal set of eigenvectors and a corresponding set of real eigenvalues; however, such an operator does *not* correspond to an observable.

leads to some surprising (i.e., nonclassical) results—among them, the so-called "wave-particle duality" described in Chapter 1.

Although Postulate 2 ascribes no real physical significance to either the observable operator $\hat{A}$ or its eigenbasis $\{\alpha_i(x)\}$, the assertion is made in part (b) that its eigenvalues $\{A_i\}$ are the *only* numbers that can be obtained in any measurement of $\mathcal{A}$. The fact that these eigenvalues are real corresponds to the fact that the measurement operation always yields a real number. However, notice that nothing in the postulate requires that these eigenvalues form a "continuous set" (i.e., that they cover densely all or part of the real number axis); indeed, our labeling of the eigenvalues seems to suggest that they form a "discrete set" (i.e., that the difference between any two given eigenvalues is some finite, nonzero number). Now as a matter of fact, the set $\{A_i\}$ may be continuous *or* discrete *or* a combination of the two, depending entirely on the particular observable operator to which the eigenvalues belong. Nevertheless, it is highly significant that *the eigenvalues can be discretely distributed*; for this immediately opens up the possibility for allowing certain physical observables to be "quantized." We recall from Chapter 1 that it was the experimental discovery of such quantized observables that formed one of the great stumbling blocks for classical mechanics.

Until now, the main simplifying restriction imposed on our development of quantum mechanics has been the restriction to systems with only *one* degree of freedom (hence the appearance of the *single* variable x as the argument of the $\mathcal{H}$-functions). We shall now impose a second simplifying restriction: henceforth, we shall treat the general observable operator $\hat{A}$ as though its eigenvalues were *entirely discretely distributed.* We do this because a discussion of operators with continuously distributed eigenvalues involves some extraordinary mathematical manipulations, which, at this point, would tend to confuse rather than enlighten. By concentrating on operators with discretely distributed eigenvalues, we shall be able to present most of the essential points of the theory with relative clarity and simplicity; moreover, many aspects of the properties of observable operators with continuously distributed eigenvalues may be understood by drawing analogies with the results for the discrete case. A brief discussion of the mathematical techniques required to deal with continuous eigenvalues will be given later in Sec. 4-6b.

In classical mechanics, if $\mathcal{A}$ is an observable then any real function of $\mathcal{A}$, $f(\mathcal{A})$, is also an observable (e.g., $\mathcal{A}^2$ or $e^{\mathcal{A}}$); this is because if we measure a value for $\mathcal{A}$, then we have obviously measured a value for $f(\mathcal{A})$ also. Does this fact carry over into quantum mechanics? The answer to this question is yes, provided we restrict ourselves to real functions $f(z)$ which have a power series expansion (i.e., a Taylor expansion) in z:

$$f(z) = \sum_{n=0}^{\infty} c_n z^n, \qquad \{c_n\} \text{ real} \tag{4-5a}$$

To see how this comes about, let us consider the *operator* $f(\hat{A})$, which is *defined* by formally replacing the variable z by the observable operator $\hat{A}$ in the series expression for $f(z)$:

$$f(\hat{A}) \equiv \sum_{n=0}^{\infty} c_n \hat{A}^n \tag{4-5b}$$

By $\hat{A}^n$, we mean of course $\hat{A}$ multiplied by itself n times; thus, in view of the definitions in Eqs. (2-42), $f(\hat{A})$ is a well-defined operator. We shall now show that $f(\hat{A})$ is in fact an *observable* operator, and that the particular observable to which it corresponds is $f(\mathfrak{A})$. These conclusions are arrived at from the results of the following exercise.

Exercise 26.

(a) Prove that the operator $c\hat{A}^n$, where $n = 0,1,2,\ldots$, has eigenvectors $\{\alpha_i(x)\}$ and eigenvalues $\{cA_i^n\}$; here, $\{\alpha_i(x)\}$ and $\{A_i\}$ are the eigenvectors and eigenvalues respectively of the observable operator $\hat{A}$. [*Hint*: Establish the result for $n = 0$ and $n = 1$; then show that the result holds for any $n \geqq 2$ provided it holds for $n - 1$.]

(b) Using the result of part (a), prove that the operator $f(\hat{A})$ has eigenvectors $\{\alpha_i(x)\}$ and eigenvalues $\{f(A_i)\}$.

We see then that the operator $f(\hat{A})$ has a complete orthonormal set of eigenvectors and a corresponding set of real eigenvalues; thus, by Postulate 2, $f(\hat{A})$ is indeed an *observable* operator. Furthermore, since the eigenvalues of $f(\hat{A})$ are $\{f(A_i)\}$, it seems quite reasonable to associate this observable operator with the particular observable $f(\mathfrak{A})$. This establishes what we wanted to prove, and it also shows that the observable operator $f(\hat{A})$ has the same eigenbasis as $\hat{A}$ does.

In classical mechanics the fact that we identify the state of a system with certain physical observables means that *both* the state and the observables depend upon time. However, in quantum mechanics no such identification is drawn between the state and the observables, and an examination of Postulates 1 and 2 leads us to conclude that, while the state vector $\Psi_t(x)$ generally changes with time, the observable operator $\hat{A}$, along with its eigenbasis $\{\alpha_i(x)\}$ and its eigenvalues $\{A_i\}$, are all *independent* of time. One consequence of this is that, if we express the state vector $\Psi_t(x)$ as a linear combination of the eigenvectors $\{\alpha_i(x)\}$ in the manner of Eq. (4-2b),

$$\Psi_t(x) = \sum_{n=1}^{\infty} (\alpha_n, \Psi_t)\alpha_n(x) \qquad \text{(4-6a)}$$

then the expansion coefficients or components of $\Psi_t(x)$ relative to the eigenbasis of Â will be *time-dependent* scalars:

$$(\alpha_n, \Psi_t) \equiv \int_{-\infty}^{\infty} \alpha_n^*(x)\Psi(x,t)\,dx \qquad (n = 1,2, \ldots) \qquad \text{(4-6b)}$$

We refer to Eq. (4-6a) as "an expansion of the state vector $\Psi_t(x)$ in the eigenbasis of the observable operator Â." At the moment, we have no apparent reason for ever wanting to write the state vector in such a way; however, as we shall see later, expansions of this type play a very key role in the quantum theory.

It should be remarked that the time-*in*dependence of Â, $\{\alpha_i(x)\}$ and $\{A_i\}$ does *not* necessarily mean that the measured values of $\mathfrak{A}$ will be constant in time. For it remains to be specified how the outcome of a particular measurement depends upon the state of the system at the time of the measurement; thus, owing to the time evolution of $\Psi_t(x)$, different eigenvalues of Â might be measured at different times. This matter will be clarified later.

4-3 THE QUANTUM THEORY OF MEASUREMENT

In Postulate 1 we have associated physical *states* with Hilbert space *vectors*, and in Postulate 2 we have associated physical *observables* with Hilbert space *operators*. However, we certainly cannot expect to form a meaningful physical theory merely by associating two separate physical entities with two separate mathematical entities. Clearly it is necessary to establish some sort of logical connection between the state vector of the system and the observable operators of the system. This logical connection is made, although in a somewhat indirect way, via the concept of *measurement*. The quantum theory of measurement thus forms the keystone of the theoretical structure of quantum mechanics; it is here that the respective roles of the state vector and the observable operators are clarified and interrelated.

It is perhaps appropriate to say again just what we mean by "a measurement": We regard a measurement simply as some in-principle well-defined physical operation which, when performed on a system, yields a single, errorless, real number. By "errorless" we mean that there is no experimental uncertainty associated with the number obtained, so that it can be regarded as being infinitely precise. Clearly what we are really contemplating here is an ideal measurement—a highly simplified abstraction of what actually occurs in the laboratory. A more thoroughgoing inquiry into the nature of a physical measurement is necessary to form a truly complete picture of the quantum theory, but such a critique is quite involved and will not be attempted here.

We mentioned in Chapter 1 that the logical foundations of quantum mechanics have been and continue to be the subject of serious debate among some physicists and philosophers of science. Most of this debate has centered upon the theory of measurement. We shall not try to discuss here the intricate pros and cons of this debate; instead, we shall merely present the essential features of the quantum theory of measurement *in the form that is currently favored by most physicists.* However, it must be noted that there are a few other tenable views of the quantum theory of measurement, all of them conflicting in various degrees with the "orthodox" view. All these views, including the orthodox one, are based on various philosophical predilections which go somewhat *beyond* what the experimental evidence directly implies; indeed, the conflict between these views is confined mainly to the philosophical level, since as yet no real laboratory experiment has been devised which is capable of picking out the "correct" view. Conversely, until such an experiment comes along, the question of which view is really more legitimate has no *practical* import.

We begin in Sec. 4-3a by stating and discussing Postulate 3, which tells us exactly what can be predicted about the outcome of a measurement of an observable $\mathcal{A}$ which is performed on a system in a known state $\Psi_t(x)$. Next, in Sec. 4-3b, we present and discuss Postulate 4, which tells us how the state vector is affected by such a measurement. Finally, in Sec. 4-3c, we derive two very important consequences of these postulates—namely, the Compatibility Theorem and the celebrated Heisenberg Uncertainty Principle.

4-3a Predicting the Result of a Measurement Expectation Values and Uncertainties

We learned in Postulate 2 that the only values that can ever be measured for an observable are the eigenvalues of the corresponding observable operator. As to which one of these eigenvalues will be obtained in any given instance, we may expect that this will be somehow determined by the particular form of the state vector at the time of the measurement. As we shall now see, the form of the state vector does indeed have an important bearing on which eigenvalue is measured, but it is in general *not possible* to predict with *absolute certainty* the outcome of a single measurement.

> **Postulate 3.** If an observable operator $\hat{A}$ has eigenbasis $\{\alpha_i(x)\}$ and eigenvalues $\{A_i\}$, and if the corresponding observable $\mathcal{A}$ is measured on a system which, immediately prior to the measurement, is in the state $\Psi_t(x)$, then the *strongest predictive statement that can be made* concerning the result of this measurement is as follows: The *probability* that the measurement will yield the eigenvalue A_k is $|(\alpha_k,\Psi_t)|^2$.†

This postulate unquestionably marks the point at which the theory of quantum mechanics diverges most radically from the theory of classical mechanics. We recall that, in classical mechanics, if the instantaneous state of the system $[x(t),p(t)]$ is known, then it is *certain* that a measurement of some observable $\mathcal{A} = f(x,p)$ at time t will yield the number $f(x(t),p(t))$. In contrast to this, Postulate 3 asserts that if the instantaneous state vector of the system $\Psi_t(x)$ is known, then all that can be predicted about a measurement of $\mathcal{A}$ at time t is that $|(\alpha_1,\Psi_t)|^2$ is the probability that the number A_1 will be obtained, $|(\alpha_2,\Psi_t)|^2$ is the probability that the number A_2 will be obtained, and so on for the other eigenvalues of A. Now since Postulate 1 asserts that the state of a system is *completely defined* by the state vector $\Psi_t(x)$, and since Postulate 3 further asserts that a knowledge of the state vector suffices *only* to predict *probabilities* for obtaining various results in a measurement, then we are forced to conclude:

(i) It is often *not* possible to predict with certainty the outcome of a measurement which is performed on a system in a *completely defined state.*

(ii) If a system is subjected to two separate but identical measurements, with due care taken to insure that the system is in the *exact same state* just prior to each measurement, the results of the two measurements will *not necessarily coincide.*

In accepting Postulate 3, it is evidently incumbent upon us also to accept this "unpredictability" and "nonuniqueness" of the measurement process as being manifestations of some inherent property

†If the eigenvalues of $\hat{A}$ were *continuously* distributed, we would have to state this postulate a bit differently. We shall discuss this point later in Sec. 4-6b.

of Nature. Although this view disagrees violently with our own deeply ingrained "classical" intuition, the most we can justifiably claim in rebuttal is that, on the *macroscopic* level, this property of Nature must not be *noticeable*. We shall return to this point later in Sec. 4-5c.

Having acknowledged this negative aspect of Postulate 3, namely that a unique result of measuring $\mathfrak{A}$ on a system in a known state $\Psi_t(x)$ usually *cannot* be predicted, let us consider now its positive aspect: the *probability* for obtaining the eigenvalue A_k *can* be predicted, and is in fact equal to $|(\alpha_k,\Psi_t)|^2$.

Exercise 27. Prove that, as required by Postulate 1, the quantity $|(\alpha_k,\Psi_t)|^2$ is not changed if $\Psi_t(x)$ is replaced by the vector $c\Psi_t(x)$, where c is any $\mathcal{H}$-scalar satisfying $|c|^2 = 1$.

We recall from Eq. (4-6a) that the inner product (α_k,Ψ_t) is just the "component" of the state vector $\Psi_t(x)$ in the "direction" of the eigenbasis vector $\alpha_k(x)$; the value of this time-dependent complex number can be calculated from the functions $\alpha_k(x)$ and $\Psi_t(x)$ according to Eq. (4-6b). It is altogether fitting that the probability for measuring A_k for $\mathfrak{A}$ in the state $\Psi_t(x)$ should be determined by the inner product (α_k,Ψ_t), since this quantity depends both upon the state vector of the system and upon some property of $\hat{A}$ associated with the eigenvalue in question. However, if the square modulus of (α_k,Ψ_t) is to be a *probability*, then we must have, in analogy with Eq. (2-2a),

$$0 \leqq |(\alpha_k,\Psi_t)|^2 \leqq 1 \qquad \text{for all } k \tag{4-7a}$$

Exercise 28. Prove the above inequality. [*Hint*: Use the Schwarz inequality to prove the right-hand relation.]

In addition to satisfying Eq. (4-7a), the inner product (α_k,Ψ_t) must also be such that

$$\sum_{k=1}^{\infty} |(\alpha_k,\Psi_t)|^2 = 1 \tag{4-7b}$$

This relation, which is analogous to Eq. (2-2b), merely expresses the requirement of Postulate 2 that a measurement of $\mathfrak{A}$ is *certain* to yield either A_1 or A_2 or A_3 . . . . In order to prove that Eq. (4-7b) is indeed satisfied, we shall show that its left hand side is just the norm of $\Psi_t(x)$; our result then will follow from the requirement of Postulate 1 that $(\Psi_t,\Psi_t) = 1$. Using the fact that the eigenvectors $\{\alpha_i(x)\}$ form an orthonormal basis in $\mathcal{H}$, we have for any possible state vector $\Psi_t(x)$,

$$(\Psi_t,\Psi_t) = \left(\sum_{i=1}^{\infty} (\alpha_i,\Psi_t)\alpha_i, \sum_{j=1}^{\infty} (\alpha_j,\Psi_t)\alpha_j\right)$$

$$= \sum_{i=1}^{\infty} \sum_{j=1}^{\infty} (\alpha_i,\Psi_t)^* (\alpha_j,\Psi_t) (\alpha_i,\alpha_j)$$

$$= \sum_{i=1}^{\infty} \sum_{j=1}^{\infty} (\alpha_i,\Psi_t)^* (\alpha_j,\Psi_t)\delta_{ij}$$

$$= \sum_{i=1}^{\infty} (\alpha_i,\Psi_t)^* (\alpha_i,\Psi_t)$$

Thus

$$(\Psi_t,\Psi_t) = \sum_{i=1}^{\infty} |(\alpha_i,\Psi_t)|^2 \tag{4-8}$$

and Eq. (4-7b) follows at once. [Note that Eq. (4-8) is just a particular case of Eq. (2-40b).] We have proved, then, that the numbers $|(\alpha_1,\Psi_t)|^2$, $|(\alpha_2,\Psi_t)|^2$, $|(\alpha_3,\Psi_t)|^2, \ldots$ satisfy conditions (4-7), and so form a set of probability numbers $p_1, p_2, p_3, \ldots$ analogous to the set discussed in Sec. 2-1. We shall explore this analogy in some detail later in this section.

If it should happen that two of the eigenvalues of $\hat{A}$ are equal, say $A_3 = A_5 = A$, then the probability for measuring the value A in the state $\Psi_t(x)$ is, according to Eq. (2-3a), just the sum of the separate probabilities, $|(\alpha_3,\Psi_t)|^2 + |(\alpha_5,\Psi_t)|^2$. In this case, we would say in the terminology of quantum mechanics that the eigenvalue A is *degenerate*. In order to keep our discussion of quantum mechanics as simple as possible, we are going to restrict our attention to observables $\mathfrak{A}$ whose operators have distinct or *nondegenerate* eigenvalues:

$$\text{Require} \quad A_i \neq A_j \quad \text{if} \quad i \neq j \quad \text{(nondegenerate eigenvalues)} \tag{4-9}$$

It is with some reluctance that we impose this simplifying restriction on our development of the theory of quantum mechanics, because it turns out that the occurrence of observable operators with degenerate eigenvalues is rather common, and is the source of many interesting phenomena. Thus we feel obliged to return to this matter at the end of our development (Sec. 4-6c), and discuss briefly how some of our conclusions will be modified by the relaxation of Eq. (4-9).

In the next exercise we shall deduce the following very important result: A measurement of $\mathcal{A}$ on the state $\Psi_t(x)$ is *certain* to yield the eigenvalue A_k if and only if $\Psi_t(x)$ coincides with $\alpha_k(x)$. We use the term "coincide" in the loose sense allowed by Postulate 1: two *state* vectors $\Psi_1(x)$ and $\Psi_2(x)$ "coincide," or describe the same physical state, if $\Psi_1(x) = c\Psi_2(x)$, where c is any complex number satisfying $|c|^2 = 1$.

Exercise 29.

(a) Prove that, if $\Psi_t(x) = c\alpha_k(x)$, where $|c|^2 = 1$, then a measurement of $\mathcal{A}$ at time t is *certain* to yield the value A_k. [*Hint*: Calculate the quantity $|(\alpha_i, \Psi_t)|^2$.]

(b) Prove that, if a measurement of $\mathcal{A}$ at time t is *certain* to yield the value A_k, then $\Psi_t(x) = c\alpha_k(x)$, where $|c|^2 = 1$. [*Hint*: Prove first that $(\alpha_i, \Psi_t) = 0$ for $i \neq k$; then use Eq. (4-6a).]

We see, then, that a necessary and sufficient condition for a measurement of $\mathcal{A}$ to yield a unique, predictable value is that the state vector of the system coincide with some eigenvector of $\hat{A}$. This is obviously a very important and useful result. However, it should *not* be taken to imply that the physical state corresponding to the normed $\mathcal{H}$-vector $\alpha_k(x)$ is in some way more precisely defined than the physical state corresponding to some normed, *linear combination* of two or more of the vectors $\{\alpha_i(x)\}$; for, Postulate 1 stipulates that *any* normed $\mathcal{H}$-vector *completely defines* a physical state, regardless of whether or not the vector in question happens to coincide with an eigenvector of some observable operator. Moreover, while $\alpha_k(x)$ may be a "nice" state vector for predicting the result of a measurement of $\mathcal{A}$, it will *not* necessarily coincide with one of the eigenvectors $\{\beta_i(x)\}$ of some other observable operator $\hat{B}$, in which case the state vector $\alpha_k(x)$ will *not* respond "nicely" to a measurement of $\mathcal{B}$. In a sense, we are merely reemphasizing here the basic point that, in quantum mechanics, the state of a system is solely and completely determined by the functional form of its state vector, and *not*, as in classical mechanics, by the expected result of any measurement which might be performed on the system.

In Sec. 2-1 we considered the problem of randomly selecting a ball from a box of N identical balls, with n_1 of the balls bearing the number v_1, n_2 bearing the number v_2, etc., and with $\Sigma_k n_k = N$. We observed that the probability p_k that the randomly selected ball will show the number v_k is n_k/N. Suppose we let

$$\left.\begin{aligned} v_k &= A_k \\ n_k &= |(\alpha_k, \Psi_t)|^2 N \end{aligned}\right\} \qquad \text{(4-10a)}$$

We note that the second of these equalities is indeed legitimate, since Eq. (4-7b) insures that $\Sigma_k n_k = N$; in addition, all of the numbers $|(\alpha_k, \Psi_t)|^2 N$ can be made as close to integer values as desired simply by taking N large enough. The probability that a randomly selected ball will show the number A_k is evidently

$$p_k = n_k/N = |(\alpha_k, \Psi_t)|^2 \tag{4-10b}$$

which is just the probability for measuring A_k in the state $\Psi_t(x)$. Therefore, we see that this hypothetical ball-drawing experiment *simulates* the process of measuring $\mathfrak{A}$ on a system in the state $\Psi_t(x)$. The major difference is that, in the ball-drawing experiment, it is in principle possible to eliminate the unpredictability and nonuniqueness of a drawing merely by ascertaining the exact positions of all the balls in the box and drawing accordingly; however, as we have previously emphasized, there is *no* possible way of eliminating the element of uncertainty in the measurement of an observable on a given state. Another difference is that, owing to the time-dependence of the inner products (α_k, Ψ_t), the numbers $\{n_i\}$ and the probabilities $\{p_i\}$ in Eqs. (4-10) will in general change with time (but note that the values $\{v_i\} = \{A_i\}$ will evidently remain fixed). However, we shall postpone until Sec. 4-4 a consideration of this time-dependence, and for now continue to confine our discussion to a single instant t.

In view of the foregoing analogy between the measurement process in quantum mechanics and the ball-drawing experiment of Sec. 2-1, it is clear that we can define for a *series* of measurements an "average value" and an "rms deviation" analogous to $\langle v \rangle$ and Δv in Eqs. (2-4) and (2-5). However we recall that, in the multiple drawing procedure used to define $\langle v \rangle$ and Δv, we were careful to return to the box each ball drawn before making the next random drawing; in other words, we wanted each drawing to be made with the box of balls in the same "state." Therefore, in order to define quantities analogous to $\langle v \rangle$ and Δv for a series of measurements, we must take care to insure that the state vector of the system is the *same* for each measurement of the series. We shall not concern ourselves here with how this can be accomplished; however, we should remark that this is not a trivial requirement, because, as we shall find in Postulate 4, the state vector of the system usually suffers a drastic alteration as a result of a measurement. For this reason, it is necessary to distinguish in our subsequent discussion two different types of multiple measurements:

(i) A series of M *repeated measurements* on the state $\Psi_t(x)$ is a series of M measurements which are performed with the system always in the state $\Psi_t(x)$ just prior to each measurement.

(ii) A series of M *successive measurements* is a series of M measurements performed in rapid succession, such that the state vector of the system for the nth measurement is the state which results from the $(n\text{-}1)$th measurement.

We shall discuss the results obtained in a series of *successive* measurements after we have introduced Postulate 4. For now, we consider a series of very many *repeated* measurements of $\mathfrak{A}$ on the state $\Psi_t(x)$: We denote by the sumbol $\langle\hat{A}\rangle_t$ the *average* of the values obtained in these repeated measurements, and by $\Delta\hat{A}_t$ the *rms deviation* of these values. Therefore, putting $v_k = A_k$ and $p_k = |(\alpha_k, \Psi_t)|^2$ into the expressions for $\langle v\rangle$ and Δv in Eqs. (2-7) and (2-9), we have at once

$$\langle\hat{A}\rangle_t = \sum_{k=1}^{\infty} |(\alpha_k, \Psi_t)|^2 A_k \tag{4-11}$$

and

$$\Delta\hat{A}_t = \sqrt{\left(\sum_{k=1}^{\infty} |(\alpha_k, \Psi_t)|^2 A_k^2\right) - \left(\sum_{k=1}^{\infty} |(\alpha_k, \Psi_t)|^2 A_k\right)^2} \tag{4-12}$$

In the terminology of quantum mechanics, $\langle\hat{A}\rangle_t$ is called "the *expectation value* of $\mathfrak{A}$ in the state $\Psi_t(x)$," and $\Delta\hat{A}_t$ is called "the *uncertainty* in $\mathfrak{A}$ in the state $\Psi_t(x)$." We must be careful, though, not to read any incorrect implications into these two names. Thus, we should not necessarily "expect" to obtain the value $\langle\hat{A}\rangle_t$ in *any* measurement of $\mathfrak{A}$ on the state $\Psi_t(x)$, because $\langle\hat{A}\rangle_t$ will not necessarily coincide with one of the eigenvalues of $\hat{A}$. In addition, the "uncertainty" described by $\Delta\hat{A}_t$ is not due to a less-than-perfect measuring technique, and measured values which differ significantly from the expectation value are no less legitimate than measured values which are nearly equal to the expectation value. In the final analysis, $\langle\hat{A}\rangle_t$ and $\Delta\hat{A}_t$ are best understood by analogy with $\langle v\rangle$ and Δv in Sec. 2-1: $\langle\hat{A}\rangle_t$, being by definition the average of the values obtained in a series of very many repeated measurements of $\mathfrak{A}$ on the state $\Psi_t(x)$, is a sort of "representative number" for all these values; $\Delta\hat{A}_t$, being by definition the rms deviation of these values, provides a quantitative estimate of their "dispersion," and hence a quantitative estimate of *how adequately* these values are "represented" by the single value $\langle\hat{A}\rangle_t$.

Exercise 30. If $f(z)$ is any real function expandable in a Taylor series, show that $\langle f(\hat{A})\rangle_t$, the expectation value of $f(\mathfrak{A})$ in the state $\Psi_t(x)$, is given by

$$\langle f(\hat{A})\rangle_t = \sum_{k=1}^{\infty} |(\alpha_k, \Psi_t)|^2 f(A_k) \tag{4-13}$$

where $\{\alpha_k(x)\}$ and $\{A_k\}$ are the eigenvectors and eigenvalues respectively of the observable operator $\hat{A}$. Compare the form of this result with Eq. (2-8). [*Hint*: To prove Eq. (4-13), apply Eq. (4-11) to the operator $f(\hat{A})$ as defined in Eq. (4-5b) and discussed in Exercise 26.]

It is significant that, according to Eq. (4-11), $\langle\hat{A}\rangle_t$ is uniquely determined by the state vector $\Psi_t(x)$. That is, although we *cannot* generally predict what we might call the "value" of $\mathcal{A}$ in the state $\Psi_t(x)$, we *can* predict the "expectation value" of A in the state $\Psi_t(x)$. Expectation values of observable operators are of practical importance for the following reason: In making measurements in the laboratory, the physicist will often effect a similtaneous measurement of many identical systems (e.g., atoms), each of which is in the *same* state $\Psi_t(x)$. Clearly, if M separate but identical systems, all in the same state $\Psi_t(x)$, are each measured once, the results will be essentially the same as if one of the systems were subjected to *M repeated* measurements in that state. Thus, the expectation value is often a very useful single number to characterize an experimental result.

Equations (4-11) and (4-12) for $\langle\hat{A}\rangle_t$ and $\Delta\hat{A}_t$ are not the most convenient forms for these two quantities. In the next two exercises, we shall derive more useful expressions for the expectation value and the uncertainty.

Exercise 31. By applying Eq. (4-13) to Eq. (4-12), prove that

$$\Delta\hat{A}_t = \sqrt{\langle\hat{A}^2\rangle_t - \langle\hat{A}\rangle_t^2} \tag{4-14}$$

The foregoing expression for the uncertainty is analogous to Eq. (2-6) for Δv, and the remarks made there are applicable here, too. The significance of Eq. (4-14) is that it expresses the uncertainty completely in terms of expectation values. This renders all the more interesting the result of the following exercise.

Exercise 32. Prove that

$$\langle\hat{A}\rangle_t = (\Psi_t, \hat{A}\Psi_t) \tag{4-15}$$

[*Hint*: Expand the inner product $(\Psi_t, \hat{A}\Psi_t)$ as we expanded (Ψ_t, Ψ_t) in our derivation of Eq. (4-8), and so obtain the right-hand side of Eq. (4-11).]

According to Eq. (4-15), we can calculate the expectation value of $\mathcal{A}$ in the state $\Psi_t(x)$ by first forming the vector $\hat{A}\Psi_t(x)$, and then taking the inner product of this vector with the vector $\Psi_t(x)$:

$$\langle\hat{A}\rangle_t = \int_{-\infty}^{\infty} \Psi_t^*(x)[\hat{A}\Psi_t(x)]\,dx \tag{4-16}$$

The virtue of Eq. (4-15) (or Eq. (4-16)) as opposed to Eq. (4-11) is that the former expresses $\langle\hat{A}\rangle_t$ in terms of the operator $\hat{A}$ and the state vector $\Psi_t(x)$ *only*, and does not involve the eigenvectors and eigenvalues of $\hat{A}$. It should also be remarked that Eqs. (4-11) and (4-12) must be modified slightly to accommodate *continuously* distributed eigenvalues, whereas Eqs. (4-15) and (4-14) are valid irrespective of the mode of distribution of the eigenvalues.

It is interesting to note that the reality of $\langle\hat{A}\rangle_t$, which of course is obvious from Eq. (4-11), can be seen from Eq. (4-15) to be a direct consequence of the fact that $\hat{A}$ is Hermitian: thus, using first Eq. (2-34a) and then Eq. (4-4), we find

$$\langle\hat{A}\rangle_t^* = (\Psi_t,\hat{A}\Psi_t)^* = (\hat{A}\Psi_t,\Psi_t) = (\Psi_t,\hat{A}\Psi_t) = \langle\hat{A}\rangle_t$$

which proves that $\langle\hat{A}\rangle_t$ is real.

We can replace $\hat{A}$ in Eq. (4-15) by $f(\hat{A})$, and so obtain

$$\langle f(\hat{A})\rangle_t = (\Psi_t,f(\hat{A})\Psi_t) \tag{4-17}$$

Applying this to Eq. (4-14), we can write $\Delta\hat{A}_t$ more explicitly as

$$\Delta\hat{A}_t = \sqrt{(\Psi_t,\hat{A}^2\Psi_t) - (\Psi_t,\hat{A}\Psi_t)^2} \tag{4-18}$$

To summarize the main features of a series of *repeated* measurements, we show in Fig. 3 the sort of results which might be expected if M repeated measurements of $\mathfrak{A}$ are performed on some state $\Psi_t(x)$. On the horizontal axis, we plot the eigenvalues of $\hat{A}$, and above each eigenvalue we draw a bar whose height is equal to the number of times that eigenvalue was measured. Ideally, we expect the eigenvalue A_i to be obtained $|(\alpha_i,\Psi_t)|^2 M$ times, but the actual number of times A_i is obtained will usually differ somewhat from this number owing to the randomness involved. In Fig. 3 we have connected the points $(A_i,|(\alpha_i,\Psi_t)|^2 M)$ with a smooth curve; we call this curve "the *distribution curve* for $\mathfrak{A}$ in the state $\Psi_t(x)$." Clearly, a *complete* specification of the expected results of these repeated measurements requires a specification of all the points $(A_i,|(\alpha_i,\Psi_t)|^2 M)$—i.e., the full distribution curve. A *less detailed* but still very useful description of these results is obtained simply by specifying the values $\langle\hat{A}\rangle_t$ and $\Delta\hat{A}_t$: these two numbers evidently characterize, insofar as it is possible, the "center" and "width," respectively, of the distribution curve.

Exercise 33. Suppose $\Psi_t(x)$ coincides with the eigenvector $\alpha_k(x)$—i.e., $\Psi_t(x) = c\alpha_k(x)$ where $|c|^2 = 1$.

(a) Show that it follows from *both* Eqs. (4-13) and (4-17) that $\langle f(\hat{A})\rangle_t = f(A_k)$.

(b) Using the result of part (a), show that in this case,

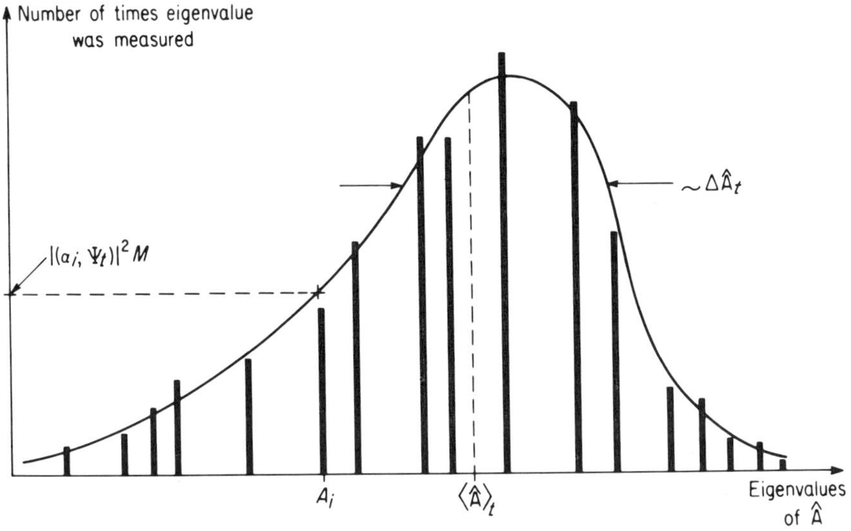

Fig. 3. A bar graph showing the eigenvalues of $\hat{A}$ versus the number of times each eigenvalue was obtained in a hypothetical series of M *repeated* measurements of $\mathfrak{A}$ on the state $\Psi_t(x)$. The smooth curve is the "distribution curve for $\mathfrak{A}$ in the state $\Psi_t(x)$," and by definition connects the points $(A_i, |(\alpha_i, \Psi_t)|^2 M)$ in a smooth but otherwise arbitrary way. The quantities $\langle\hat{A}\rangle_t$ and $\Delta\hat{A}_t$ measure the mean and rms spread of the distribution curve.

$\langle\hat{A}\rangle_t = A_k$ and $\Delta\hat{A}_t = 0$. Describe the shape of the distribution curve for $\mathfrak{A}$ in the state $\alpha_k(x)$.

4-3b The Effect of a Measurement upon the State The "Value of an Observable"

According to Postulate 3, if two *repeated* measurements of some observable $\mathfrak{A}$ are made on a system in a given state, the results of these two measurements will *not* usually coincide. If we were actually accustomed to making measurements on the microscopic level, this fact would seem commonplace to us. Equally commonplace would be the following fact: if two *successive* measurements of $\mathfrak{A}$ are made on a system, the results *would* always coincide. More specifically, suppose the system is in some state $\Psi_t(x)$, not necessarily an eigenvector of $\hat{A}$, and suppose that a measurement of $\mathfrak{A}$ yields the eigenvalue A_k. If we then make an immediate remeasurement of $\mathfrak{A}$, without "readjusting" the state vector as we did for repeated measurements, we can be certain of obtaining the same result, A_k. The remeasurement is to be made "immediately" in order to insure that no time evolution of the state vector takes place between the measurement and the remeasurement.†

The *predictability* of *successive* measurements of $\mathcal{A}$, when contrasted with the *unpredictability* of *repeated* measurements of $\mathcal{A}$, stands as a relatively "charitable gesture" on the part of Nature. To see how this feature may be incorporated into our theory of quantum mechanics, we reason as follows: According to Exercise 29, if we are *certain* that the remeasurement of $\mathcal{A}$ will yield the same eigenvalue A_k as did the measurement, then the state vector of the system at the time of the remeasurement *must* coincide with the eigenvector $\alpha_k(x)$. This implies that the state vector of the system immediately *after* the first measurement must coincide with $\alpha_k(x)$, *regardless* of what the state vector was just *before* the first measurement. These considerations make plausible (but of course are not intended to "prove") our fourth postulate, which tells us how the state vector is affected by a measurement.

> **Postulate 4.** A measurement of an observable generally causes a drastic, uncontrollable alteration in the state vector of the system; specifically, regardless of the form of the state vector just *before* the measurement, immediately *after* the measurement it will coincide with the eigenvector corresponding to the eigenvalue obtained in the measurement.‡

Exercise 34. Prove that, as a consequence of Postulates 3 and 4, two *successive* measurements of $\mathcal{A}$ will necessarily yield identical results, assuming the measurements are performed sufficiently close together in time.

Postulate 4 asserts that a measurement of an observable $\mathcal{A}$ essentially forces the state vector of the system into an eigenvector of $\hat{A}$. However, it makes no claims regarding the details of the process by which this change in the state vector occurs. Indeed, since it is generally not possible to predict with certainty which eigenvalue will be obtained in a measurement, it follows that it is generally not possible to predict with certainty which eigenvector the state of the system will be forced into by the measurement. All we can say is that the *probability* that a measurement of $\mathcal{A}$ will force the system

†The time evolution of the state vector is the subject of Postulate 5, and will be discussed in Sec. 4-4.

‡If our development allowed for observable operators with *degenerate* eigenvalues, we would have to state this postulate a bit differently. We shall discuss this point later in Sec. 4-6c.

into the eigenvector $\alpha_k(x)$ is $|(\alpha_k, \Psi_t)|^2$, where $\Psi_t(x)$ is the state vector of the system just prior to the measurement. Thus, the same element of "indeterminism" which pervaded Postulate 3 greets us again in Postulate 4.

A rather curious implication of Postulates 3 and 4 is that a measurement tells us much more about the state of the system immediately *after* the measurement than the state of the system immediately *before* the measurement. For, suppose a measurement of $\mathcal{A}$ yields the eigenvalue A_k: by Postulate 4, we can deduce that the state vector of the system immediately *after* the measurement coincides with the eigenvector $\alpha_k(x)$; however, all we can say about the state vector immediately *before* the measurement is that, by Postulate 3, its inner product with $\alpha_k(x)$ was nonzero. Thus the outcome of a single measurement tells us rather precisely what the state of the system is *as a result of our having measured it*, but very little about what the state of the system was when we *started* measuring it. In a sense, then, the measurement operation in quantum mechanics is more in the nature of a "preparation" of a state, rather than an "observation" of a state (although we evidently have no real control over which state will be "prepared" in a given measurement).

The reader should now begin to appreciate the really profound difference between the classical and quantum theories of measurement. This difference can be further illuminated by a careful examination of the meaning of the phrase "the value of an observable."

In classical mechanics, we regard an observable as *always* "having a value," and a measurement of an observable simply amounts to taking an unobtrusive peek at what its current value really is. In particular, the value of an observable is presumed to *exist* regardless of whether or not it is *perceived* in a measurement by some observer.

In quantum mechanics, for the particular case in which the state vector $\Psi_t(x)$ coincides with some eigenvector $\alpha_k(x)$ of $\hat{A}$, we can evidently adopt without difficulty the viewpoint of classical mechanics and say that $\mathcal{A}$ "has the value" A_k. The justification for this is simply that, by Postulate 3, it is absolutely certain that an ideal measurement of $\mathcal{A}$ will yield the number A_k. We run into difficulties, however, when we consider the more general case in which $\Psi_t(x)$ is a linear combination of two or more eigenvectors of $\hat{A}$:

$$\Psi_t(x) = \sum_i (\alpha_i, \Psi_t)\alpha_i(x) \tag{4-19}$$

In this case, a measurement of $\mathcal{A}$ can clearly yield *any* value A_i for which the corresponding component (α_i, Ψ_t) is nonzero; therefore the value obtained in a measurement of $\mathcal{A}$ on this state will be *neither predictable nor unique*. For this reason, it seems rather pointless, if not meaningless, to try to ascribe some particular "value" to $\mathcal{A}$ in this state. In fact, the prevailing view among physicists today is that *if the state vector of the system does not coincide with an eigenvector of* $\hat{A}$, *then the corresponding observable* $\mathcal{A}$ *cannot be said to "have a value" in the generally accepted sense of this phrase.* There are other tenable points of view on this somewhat philosophical matter; however, the foregoing is the "orthodox" view, and therefore is the one that we shall adopt in this introductory treatment of quantum mechanics.

The above conclusion is such a radical departure from our classical way of thinking that we cannot help but look for some easy way of avoiding it. For example, it is tempting to try to interpret the expansion of $\Psi_t(x)$ in Eq. (4-19) as being merely a shorthand way of saying that "the state vector of the system *really* coincides with *one particular* eigenvector $\alpha_i(x)$, but owing to a lack of information we can say only that $|(\alpha_j,\Psi_t)|^2$ is the *probability* that this particular eigenvector is $\alpha_j(x)$." Now, if this interpretation of Eq. (4-19) were legitimate, then $\mathcal{A}$ could indeed be said to "have a value"; we just wouldn't know for sure what the value really is, since we don't know for sure with which eigenvector $\Psi_t(x)$ really coincides. However, according to the orthodox view of quantum mechanics, this is *not* a correct interpretation of the expansion in Eq. (4-19). For, this linear combination is a perfectly legitimate normed $\mathcal{H}$-vector, and so by Postulate 1 defines a physical state of the system just as precisely and completely as does any one of the eigenvectors of $\hat{A}$. Moreover, if it so happened that this linear combination coincided with some eigenvector $\beta_k(x)$ of some other observable operator $\hat{B}$ (in which case $\mathcal{B}$ could be said to "have the value B_k"), then it would be *inconsistent* to say that the state vector *really* coincides with one of the eigenvectors $\alpha_i(x)$. We see, then, that within the framework of the four postulates which we have laid down here, there seems to be no simple, satisfactory way around the conclusion that an observable does *not always* "have a value."

Generally speaking, then, a measurement of $\mathcal{A}$ in quantum mechanics is *not* simply a matter of "taking an unobtrusive peek at the value of $\mathcal{A}$." For, if the state vector of the system does *not* coincide with one of the eigenvectors of $\hat{A}$, then (i) the measurement cannot be "unobtrusive," since the state vector will necessarily be altered by the measurement, and (ii) a "value of $\mathcal{A}$" does not even exist in the usual sense of the phrase, since the result of the measurement is not uniquely predetermined. Perhaps a more accurate description of the measurement process is to say that *the very act of measuring $\mathcal{A}$ essentially develops a value of $\mathcal{A}$*; it evidently accomplishes this by the simple expedient of forcing the state vector into one of the eigenvectors of $\hat{A}$, so that *then* $\mathcal{A}$ will indeed have a value. In this view, $|(\alpha_k, \Psi_t)|^2$ is the probability that a measurement of $\mathcal{A}$ on the state $\Psi_t(x)$ will *develop* the value A_k. But, to repeat ourselves, this is *not* the same as saying that $|(\alpha_k, \Psi_t)|^2$ is the probability that $\mathcal{A}$ *has* the value A_k in the state $\Psi_t(x)$. Regardless of one's point of view on this somewhat philosophical matter, it clearly is usually safer to speak of the "value obtained in a measurement of $\mathcal{A}$" rather than the "value of $\mathcal{A}$."

We shall agree, then, that it is *strictly* legitimate to say that $\mathcal{A}$ "has a value" in the state $\Psi_t(x)$ if and only if a measurement of $\mathcal{A}$ on this state is certain to yield a definite result—i.e., if and only if $\Psi_t(x)$ coincides with an eigenvector of $\hat{A}$. Suppose it happens that $\Psi_t(x)$ "almost" coincides with an eigenvector of $\hat{A}$; more specifically, suppose that in the expansion of $\Psi_t(x)$ in the eigenbasis of $\hat{A}$, [Eq. (4-19)], the coefficient of one particular eigenvector $\alpha_k(x)$ strongly dominates the coefficients of all the other eigenvectors, so that in accordance with Eq. (4-7b),

$$|(\alpha_i, \Psi_t)|^2 \begin{cases} <<1 & \text{for } i \neq k \\ \lesssim 1 & \text{for } i \neq k \end{cases}$$

Exercise 35. If $\Psi_t(x)$ is such that the above conditions are satisfied prove that:

(a) The distribution curve [see Fig. 3] for $\mathcal{A}$ in this state has a sharp, narrow peak at the eigenvalue A_k.

(b) A measurement of $\mathcal{A}$ on this state is *almost* certain to yield a definite result.

In view of conclusion (b) in the above exercise, it is tempting to say that $\mathcal{A}$ *almost* has a value in the state $\Psi_t(x)$. The strength of the qualifier "almost" is evidently governed by the width of the distribution curve for $\mathcal{A}$ in this state. Since this width is proportional to $\Delta\hat{A}_t$, we may therefore regard the uncertainty in $\mathcal{A}$ in the state $\Psi_t(x)$ as indicating the extent to which $\mathcal{A}$ can be said to "have a value" in this state: the *smaller* $\Delta\hat{A}_t$ is, the *more* sense it makes to say $\mathcal{A}$ "has a value," and the *larger* $\Delta\hat{A}_t$ is, the *less* sense it makes to say this. This rather loose way of speaking will prove useful in interpreting some of our subsequent results.

We shall conclude our general discussion of the quantum theory of measurement in the next section, where we shall derive two important theorems pertaining to the measurement of *two* observables, $\mathcal{A}$ and $\mathcal{B}$.

4-3c The Compatibility Theorem and the Heisenberg Uncertainty Principle

We come now to consider one of the most interesting and important topics in quantum mechanics, namely, the problem of the "simultaneous measurability" or "compatibility" of two observables. Let us begin by explaining precisely what we mean by these terms.

Suppose a given system is subjected to three *successive* measurements involving two observables, $\mathcal{A}$ and $\mathcal{B}$: the first measurement, denoted by $M_{\mathcal{A}}$, measures $\mathcal{A}$; the second measurement, denoted by $M_{\mathcal{B}}$, measures $\mathcal{B}$; and finally the third measurement, denoted by $M'_{\mathcal{A}}$, measures $\mathcal{A}$ again. It is of course understood that these measurements are to be performed in very rapid succession so that there is no significant time evolution of the state of the system between $M_{\mathcal{A}}$ and $M_{\mathcal{B}}$, and between $M_{\mathcal{B}}$ and $M'_{\mathcal{A}}$. With respect to these measurements, we now make the following definition: the observables $\mathcal{A}$ and $\mathcal{B}$ are said to be *simultaneously measurable* or *compatible* if and only if the result of $M'_{\mathcal{A}}$ is *certain* to coincide with the result of $M_{\mathcal{A}}$, regardless of what the state of the system was just prior to $M_{\mathcal{A}}$.

From the standpoint of classical mechanics this definition is rather useless, because in classical mechanics *all* pairs of observables are "compatible": $\mathcal{A}$ and $\mathcal{B}$ each "have a value" at all times, and since the (ideal) measurement $M_{\mathcal{B}}$ will have no effect upon the value of $\mathcal{A}$, it follows that $M_{\mathcal{A}}$ and $M'_{\mathcal{A}}$ will always yield the same number.

From the standpoint of quantum mechanics, though, we can easily see that $\mathcal{A}$ and $\mathcal{B}$ might very well fail to satisfy our condition for being simultaneously measurable: We denote by $\hat{A}$ and $\hat{B}$ the observable operators for $\mathcal{A}$ and $\mathcal{B}$, and we denote by $\{\alpha_i(x)\}$, $\{\beta_i(x)\}$ and $\{A_i\}$, $\{B_i\}$ the associated eigenbases and eigenvalues of these operators.

$$\left.\begin{aligned} \hat{A}\alpha_i(x) &= A_i\alpha_i(x) \quad i = 1,2,\ldots \\ \hat{B}\beta_i(x) &= B_i\beta_i(x) \quad i = 1,2,\ldots \end{aligned}\right\} \tag{4-20}$$

Suppose that $M_{\mathcal{A}}$ yields the value A_n, and $M_{\mathcal{B}}$ then yields the value B_m. By Postulate 4, the state vector just after $M_{\mathcal{B}}$, or just before $M'_{\mathcal{A}}$, coincides with $\beta_m(x)$. Now, it is not necessarily the case that $\beta_m(x)$

coincides with $\alpha_n(x)$; if it does not, then by Postulate 3 (or Exercise 29) $M'_{\mathcal{A}}$ will not *necessarily* yield the value A_n, and therefore $\mathcal{A}$ and $\mathcal{B}$ are not "compatible."

If, in the foregoing example, the second measurement $M_{\mathcal{B}}$ had not been performed, then the results of $M_{\mathcal{A}}$ and $M'_{\mathcal{A}}$ would obviously have been the same [see Exercise 34]. This implies that $M_{\mathcal{B}}$ always has the potential of "spoiling" the remeasurement, $M'_{\mathcal{A}}$. It must be emphasized that this spoilage, if it occurs, is *not* the result of an imperfect measuring technique, but rather follows as a simple, direct consequence of Postulates 3 and 4.

We shall now state and prove one of the fundamental theorems of quantum mechanics, which we shall call the "Compatibility Theorem." This theorem essentially provides us with *two* conditions, *either* of which is both necessary and sufficient for $\mathcal{A}$ and $\mathcal{B}$ to be compatible observables. The proof of this theorem is in itself an illuminating exercise in the application of Postulates 3 and 4. Our proof here will be subject to our continuing restriction that $\hat{A}$ and $\hat{B}$ have nondegenerate eigenvalues [see Eq. (4-9)]; the theorem is actually valid without this restriction, but certain parts of the proof are more complicated.†

The Compatibility Theorem. Given two observables $\mathcal{A}$ and $\mathcal{B}$ with corresponding operators $\hat{A}$ and $\hat{B}$, then any one of the following three conditions implies the other two:

(i) $\mathcal{A}$ and $\mathcal{B}$ are compatible observables.
(ii) $\hat{A}$ and $\hat{B}$ possess a common eigenbasis.
(iii) $\hat{A}$ and $\hat{B}$ commute.

Proof: Our proof will consist in showing that (i) and (ii) imply each other, and (ii) and (iii) imply each other. The fact that (i) and (iii) imply each other will then follow trivially.

(i) *implies* (ii): Suppose that, just prior to $M_{\mathcal{A}}$, the state vector of the system coincides with *any* eigenvector $\alpha_i(x)$ of $\hat{A}$. Then $M_{\mathcal{A}}$ will yield the value A_i [by Exercise 29]. When $M_{\mathcal{B}}$ is performed, the state vector will become coincident with *some* eigenvector $\beta_j(x)$ of $\hat{B}$ [by Postulate 4]. Now since $\mathcal{A}$ and $\mathcal{B}$ are given to be compatible, then it is certain that the third measurement $M'_{\mathcal{A}}$ must yield the same value A_i that was obtained in the first measurement. But by Exercise 29, if a measurement of $\mathcal{A}$ performed on the state $\beta_j(x)$ is certain to yield the eigenvalue A_i, then $\beta_j(x)$ must coincide with $\alpha_i(x)$. We have proved then that *any* vector of the set $\{\alpha_i(x)\}$ coincides with *some*

†For a general proof of the Compatibility Theorem, see Chapter IV of F. Mandl, *Quantum Mechanics*, cited in footnote in Preface.

vector of the set $\{\beta_j(x)\}$. Since these sets are orthonormal basis sets, the correspondence must be one-to-one;† thus, simply by rearranging indices we can put $\{\alpha_n(x)\} = \{\beta_n(x)\} = \{\phi_n(x)\}$, where $\{\phi_n(x)\}$ is a "common eigenbasis" for Â and B̂:

$$\left.\begin{aligned} \hat{A}\phi_n(x) &= A_n\phi_n(x) \qquad n = 1,2,\ldots \\ \hat{B}\phi_n(x) &= B_n\phi_n(x) \qquad n = 1,2,\ldots \end{aligned}\right\} \tag{4-21}$$

(ii) *implies* (i): Given the common eigenbasis $\{\phi_n(x)\}$ as in Eqs. (4-21), measurement $M_{\mathcal{A}}$, with any result A_n, leaves the system in the state $\phi_n(x)$, by Postulate 4. Then by Exercise 29, $M_{\mathcal{B}}$ must yield the value B_n, and what is more important, will *leave* the system in the state $\phi_n(x)$. Hence, by Exercise 29, $M'_{\mathcal{A}}$ must yield the value A_n again. Since these arguments are independent of which eigenvalue A_n was obtained in the first measurement, it therefore holds regardless of what state the system was in just prior to that measurement; consequently, $\mathcal{A}$ and $\mathcal{B}$ are compatible observables.

(ii) *implies* (iii): Given the common eigenbasis $\{\phi_n(x)\}$ as in Eqs. (4-21), then using the linearity of Â and B̂ we have

$$\hat{A}\hat{B}\phi_n(x) = \hat{A}B_n\phi_n(x) = B_n\hat{A}\phi_n(x) = B_nA_n\phi_n(x)$$

$$\hat{B}\hat{A}\phi_n(x) = \hat{B}A_n\phi_n(x) = A_n\hat{B}\phi_n(x) = A_nB_n\phi_n(x)$$

Thus, $(\hat{A}\hat{B} - \hat{B}\hat{A})\phi_n(x) = 0$. Now we are not yet done, for in order to show that Â and B̂ commute, we must show that $(\hat{A}\hat{B} - \hat{B}\hat{A})\psi(x) = 0$ for *any* $\mathcal{H}$-vector $\psi(x)$. To this end, we first expand the given $\mathcal{H}$-vector $\psi(x)$ in the eigenbasis $\{\phi_n(x)\}$:

$$\psi(x) = \sum_n c_n\phi_n(x)$$

where $c_n \equiv (\phi_n, \psi)$ [see Eq. (2-39)]. Then, using the fact that both the product and the sum of two linear operators are also linear opera-

† This is "almost obvious." We have shown that *each* α-vector coincides with *some* β-vector. Now, two α-vectors cannot coincide with the *same* β-vector, since any two α-vectors are orthogonal; similarly, two β-vectors cannot coincide with the same α-vector. It remains only to show that no β-vector is "missed" in the α-to-β correspondence. To prove this, assume the contrary: some β-vector, $\beta_k(x)$, does not coincide with any of the α-vectors. Then the expansion of $\beta_k(x)$ in the α-basis, $\beta_k(x) = \Sigma_i(\alpha_i,\beta_k)\alpha_i(x)$, must contain *at least two* nonvanishing terms, thus implying that $\beta_k(x)$ is nonorthogonal to at least two α-vectors. But since each α-vector coincides with a unique β-vector, then we must conclude that $\beta_k(x)$ is nonorthogonal to at least two β-vectors—a conclusion which clearly contradicts the orthonormality of the β-vectors. Therefore the assumption is false, and the α-β correspondence is indeed one-to-one.

tors, we find

$$(\hat{A}\hat{B} - \hat{B}\hat{A})\psi(x) = \sum_n c_n(\hat{A}\hat{B} - \hat{B}\hat{A})\phi_n(x) = \sum_n c_n \cdot 0 = 0$$

Thus the operator $\hat{A}\hat{B}$ acting on any $\mathcal{H}$-vector $\psi(x)$ produces the same vector as does the operator $\hat{B}\hat{A}$ acting on $\psi(x)$. This means that $\hat{A}$ and $\hat{B}$ commute.

(iii) *implies* (ii): Given that $\hat{A}$ and $\hat{B}$ commute, then for *any* eigenvector $\alpha_i(x)$ of $\hat{A}$, we have

$$\hat{A}\hat{B}\alpha_i(x) = \hat{B}\hat{A}\alpha_i(x) = \hat{B}A_i\alpha_i(x) = A_i\hat{B}\alpha_i(x)$$

In other words, $\hat{A}$ operating on the $\mathcal{H}$-vector $B\alpha_i(x)$ has the effect of simply multiplying this vector by the number A_i; this implies that the vector $\hat{B}\alpha_i(x)$ is an eigenvector of $\hat{A}$ belonging to the eigenvalue A_i. Since the eigenvalues of $\hat{A}$ are taken to be nondegenerate, then the vector $\hat{B}\alpha_i(x)$ can differ from the vector $\alpha_i(x)$ by at most a scalar factor c, which merely takes account of the fact that the norms of $\alpha_i(x)$ and $\hat{B}\alpha_i(x)$ need not be equal:

$$\hat{B}\alpha_i(x) = c\alpha_i(x)$$

But this equation implies that $\alpha_i(x)$ is a (normed) eigenvector of $\hat{B}$ belonging to the eigenvalue c. Since the eigenvalues of $\hat{B}$ are nondegenerate, then we have for some j, $c = B_j$ and $\alpha_i(x) = \beta_j(x)$. We have proved then that *any* vector of the set $\{\alpha_i(x)\}$ coincides with *some* vector of the set $\{\beta_j(x)\}$. Since these sets are orthonormal basis sets, the correspondence must be one-to-one,† so simply by rearranging indices we can put $\{\alpha_n(x)\} = \{\beta_n(x)\} = \{\phi_n(x)\}$, where $\{\phi_n(x)\}$ is the "common eigenbasis" set in Eq. (4-21).

Q.E.D.

As a simple illustration of this theorem, we recall from Exercise 26 that, for any reasonable function f, the operator $f(\hat{A})$ is an observable operator which has the same eigenbasis as the observable operator $\hat{A}$. Therefore, the Compatibility Theorem tells us that the operators $\hat{A}$ and $f(\hat{A})$ must *commute*, as may easily be verified from Eq. (4-5b), and also that the observables $\mathcal{A}$ and $f(\mathcal{A})$ must be *compatible*, which seems eminently reasonable.

Exercise 36.

(a) Using the power series expansion of $f(\hat{A})$ in Eq. (4-5b), prove directly that $\hat{A}$ and $f(\hat{A})$ commute.

(b) Using the fact that $\hat{A}$ and $f(\hat{A})$ have the same eigenbasis $\{\alpha_i(x)\}$, prove directly that $\mathcal{A}$ and $f(\mathcal{A})$ are compatible.

†See preceding footnote.

If Â and B̂ do *not* possess a common eigenbasis, then according to the Compatibility Theorem two successive measurements $M_{\mathfrak{a}}$ and $M'_{\mathfrak{a}}$ of $\mathfrak{a}$, when separated by a measurement $M_{\mathfrak{B}}$ of $\mathfrak{B}$, will *not always* yield identical results. However, for a given result of $M_{\mathfrak{a}}$, it *is* possible to make *probabilistic* predictions about the outcome of $M'_{\mathfrak{a}}$. To do this, one first writes down the expansion of the Â-eigenbasis in terms of the B̂-eigenbasis and vice versa:

$$\alpha_k(x) = \sum_{i=1}^{\infty} (\beta_i, \alpha_k)\beta_i(x)$$

$$\beta_k(x) = \sum_{i=1}^{\infty} (\alpha_i, \beta_k)\alpha_i(x)$$

Using these expansions, along with Postulates 3 and 4, one can calculate the probabilities for all possible "routes" which lead from the given result of $M_{\mathfrak{a}}$ to any particular result of $M'_{\mathfrak{a}}$. Thus, suppose $M_{\mathfrak{a}}$ yields A_n, and we wish to know the probability that $M'_{\mathfrak{a}}$ will yield A_m. We reason as follows: After $M_{\mathfrak{a}}$ yields A_n, the system is in the state $\alpha_n(x)$, and the probability that $M_{\mathfrak{B}}$ will yield some result B_i is $|(\beta_i, \alpha_n)|^2$. If B_i is obtained on the second measurement, the system will be in the state $\beta_i(x)$ for $M'_{\mathfrak{a}}$, so the probability that $M'_{\mathfrak{a}}$ will then yield the desired result A_m is $|(\alpha_m, \beta_i)|^2$. Thus, according to Eq. (2-3b), the product $|(\beta_i, \alpha_n)|^2 \, |(\alpha_m, \beta_i)|^2$ gives the probability that the value A_m will be obtained in $M'_{\mathfrak{a}}$ *via* the result B_i for $M_{\mathfrak{B}}$. Using Eq. (2-3a), we conclude that the sum $\Sigma_i |(\beta_i, \alpha_n)|^2 \, |(\alpha_m, \beta_i)|^2$ is the probability that, given the result A_n of $M_{\mathfrak{a}}$, $M'_{\mathfrak{a}}$ will yield the value A_m *regardless* of the result of $M_{\mathfrak{B}}$. The following exercise will illustrate this procedure more explicitly.

Exercise 37. Suppose Â and B̂ "almost" possess a common eigenbasis; more specifically, suppose that when the eigenvectors of B̂ are expanded in terms of the eigenvectors of Â, one has

$$\left.\begin{aligned} \beta_1(x) &= \frac{\sqrt{3}}{2}\alpha_1(x) + \frac{1}{2}\alpha_2(x) \\ \beta_2(x) &= \frac{1}{2}\alpha_1(x) - \frac{\sqrt{3}}{2}\alpha_2(x) \\ \beta_n(x) &= \alpha_n(x), \qquad n \geqq 3 \end{aligned}\right\}$$

(a) Verify that this expansion is consistent with the orthornormality of $\{\alpha_i(x)\}$ and $\{\beta_i(x)\}$; i.e., prove that if $(\alpha_i, \alpha_j) = \delta_{ij}$, then $(\beta_i, \beta_j) = \delta_{ij}$ also.

(b) Expand the eigenvectors of $\hat{A}$ in terms of the eigenvectors of $\hat{B}$.
(c) Prove that if $M_{\mathcal{A}}$ yields any one of the values $A_3, A_4, \ldots,$ then $M'_{\mathcal{A}}$ will necessarily yield the same result.
(d) Prove that if $M_{\mathcal{A}}$ yields the value A_1, then there is a 5/8 probability that $M'_{\mathcal{A}}$ will yield A_1 and a 3/8 probability that $M'_{\mathcal{A}}$ will yield A_2.

If two observables $\mathcal{A}$ and $\mathcal{B}$ are incompatible, then the Compatibility Theorem tells us that their corresponding operators $\hat{A}$ and $\hat{B}$ do not commute; that is, for at least one vector $\psi(x)$ in $\mathcal{H}$, $(\hat{A}\hat{B} - \hat{B}\hat{A})\,\psi(x) \neq 0$. Now it turns out that the noncommutability of many pairs of noncommuting observable operators can be expressed by an equation of the form

$$\hat{A}\hat{B} - \hat{B}\hat{A} = c \tag{4-22}$$

where c is some nonzero scalar; Eq. (4-22) means simply that, for *any* $\mathcal{H}$-vector $\psi(x)$, $(\hat{A}\hat{B} - \hat{B}\hat{A})\psi(x) = c\psi(x)$. For such a case, the inherent incompatibility of $\mathcal{A}$ and $\mathcal{B}$ is strikingly illustrated by the famous "Uncertainty Principle," which was first ennunciated by W. Heisenberg:

The Heisenberg Uncertainty Principle. If $\hat{A}$ and $\hat{B}$ are such that $\hat{A}\hat{B} - \hat{B}\hat{A} = c$, where c is a scalar, then the uncertainties in $\mathcal{A}$ and $\mathcal{B}$ in *any* state $\Psi_t(x)$ satisfy

$$\Delta\hat{A}_t \cdot \Delta\hat{B}_t \geqq \frac{1}{2}\,|c| \tag{4-23}$$

Before presenting the proof of this theorem (the proof is purely mathematical and involves no physical arguments), let us first point out its profound physical implications.

According to the Heisenberg Uncertainty Principle, if two observable operators $\hat{A}$ and $\hat{B}$ satisfy Eq. (4-22) with $c \neq 0$, then the product of the uncertainties in $\mathcal{A}$ and $\mathcal{B}$ in any state $\Psi_t(x)$ is strictly bounded away from zero. Thus, if we somehow contrive to force the system into states having smaller and smaller uncertainties in $\mathcal{A}$, then these states will necessarily have larger and larger uncertainties in $\mathcal{B}$—and vice versa. In view of our discussion at the end of Sec. 4-3b, we can also express these conclusions in the following way: If $\hat{A}$ and $\hat{B}$ satisfy Eq. (4-22) with $c \neq 0$, then Eq. (4-23) implies that the *more* sense it makes to say that $\mathcal{A}$ "has a value" in a given state, the *less* sense it makes to say that $\mathcal{B}$ "has a value" in that state—and vice versa. Furthermore, if it so happens that $\Psi_t(x)$ coincides with one of the eigenvectors of $\hat{A}$, so that $\Delta\hat{A}_t = 0$ and $\mathcal{A}$ therefore *really* has a value, then Eq. (4-23) evidently requires that $\Delta\hat{B}_t = \infty$, in which case it would be *completely meaningless* to speak of $\mathcal{B}$ as having a value.

In view of these implications, it is not surprising that the Heisenberg Uncertainty Principle occupies a prominent place in the overall structure of quantum mechanics. For example, we shall see in Sec. 4-5c that it leads to a satisfactory resolution of the "wave-particle paradox."

We shall now conclude our development of the quantum theory of measurement by proving the Heisenberg Uncertainty Principle. Although the proof entails a considerable amount of mathematical manipulation, the reader should note that it relies mainly upon (i) the definition of Hermiticity in Eq. (4-4), (ii) the mathematical expression for the uncertainty in Eq. (4-14), and (iii) the properties of the inner product of two $\mathcal{H}$-vectors in Eqs. (2-34) (particularly the Schwarz inequality).

To simplify our notation, we shall omit the subscript t in what follows, keeping in mind the fact that all our calculations hold at any one instant of time. Using the expressions

$$\Delta\hat{A} = \sqrt{\langle\hat{A}^2\rangle - \langle\hat{A}\rangle^2} \qquad \Delta\hat{B} = \sqrt{\langle\hat{B}^2\rangle - \langle\hat{B}\rangle^2}$$

for the uncertainties in $\mathcal{A}$ and $\mathcal{B}$ in the state $\Psi(x)$, we shall first prove the *generalized uncertainty relation:*

$$\Delta\hat{A}\cdot\Delta\hat{B} \geqq \frac{1}{2}\,|(\Psi, [\hat{A}\hat{B} - \hat{B}\hat{A}]\,\Psi)| \tag{4-24}$$

From this generally valid relation, we may easily obtain Eq. (4-23) by inserting Eq. (4-22) and then using the fact that $(\Psi,\Psi) = 1$.

To prove Eq. (4-24), we first define two operators $\hat{A}'$ and $\hat{B}'$ by

$$\hat{A}' \equiv \hat{A} - \langle\hat{A}\rangle \quad \text{and} \quad \hat{B}' \equiv \hat{B} - \langle\hat{B}\rangle$$

Concerning these operators, we next prove three lemmas.

Lemma 1. $\hat{A}'$ and $\hat{B}'$ are Hermitian operators.

Exercise 38. Prove Lemma 1. [*Hint*: Using the fact that $\hat{A}'$ is the difference of two Hermitian operators, show that $(\psi_1, \hat{A}'\psi_2) = (\hat{A}'\psi_1, \psi_2)$ for any two Hilbert space vectors $\psi_1(x)$ and $\psi_2(x)$.]

Lemma 2. $\hat{A}'\hat{B}' - \hat{B}'\hat{A}' = \hat{A}\hat{B} - \hat{B}\hat{A}$

Exercise 39. Prove Lemma 2. [*Hint:* Note, for example, that $\hat{A}\langle\hat{B}\rangle = \langle\hat{B}\rangle\hat{A}$, since $\hat{A}$ is a *linear* operator and $\langle\hat{B}\rangle$ is a *scalar*.]

Lemma 3. $(\hat{A}'\Psi, \hat{A}'\Psi) = (\Delta\hat{A})^2$

The proof of Lemma 3 goes as follows:

$$\begin{aligned}(\hat{A}'\Psi,\hat{A}'\Psi) &= (\Psi,\hat{A}'^2\Psi)\\ &= (\Psi,[\hat{A}-\langle\hat{A}\rangle]^2\Psi)\\ &= (\Psi,[\hat{A}^2-2\langle\hat{A}\rangle\hat{A}+\langle\hat{A}\rangle^2]\Psi)\\ &= (\Psi,\hat{A}^2\Psi)-2\langle\hat{A}\rangle(\Psi,\hat{A}\Psi)+\langle\hat{A}\rangle^2(\Psi,\Psi)\\ &= \langle\hat{A}^2\rangle-2\langle\hat{A}\rangle\langle\hat{A}\rangle+\langle\hat{A}\rangle^2\cdot 1\\ &= \langle\hat{A}^2\rangle-\langle\hat{A}\rangle^2\end{aligned}$$

$$\therefore(\hat{A}'\Psi,\hat{A}'\Psi) = (\Delta\hat{A})^2$$

The first step in the above proof makes use of the Hermiticity of $\hat{A}'$, which was established in Lemma 1. The remaining steps invoke the definitions of $\hat{A}'$, $\langle\hat{A}\rangle$ and $\Delta\hat{A}$, as well as the mathematical properties expressed in Eqs. (2-34) and (2-42).†

With these three lemmas, the proof of the generalized uncertainty relation in Eq. (4-24) goes as follows:

$$\begin{aligned}(\Psi,[\hat{A}\hat{B}-\hat{B}\hat{A}]\Psi) &= (\Psi,[\hat{A}'\hat{B}'-\hat{B}'\hat{A}']\Psi) && \text{[by Lemma 2]}\\ &= (\Psi,\hat{A}'\hat{B}'\Psi)-(\Psi,\hat{B}'\hat{A}'\Psi)\\ &= (\hat{A}'\Psi,\hat{B}'\Psi)-(\hat{B}'\Psi,\hat{A}'\Psi) && \text{[by Lemma 1]}\\ &= (\hat{A}'\Psi,\hat{B}'\Psi)-(\hat{A}'\Psi,\hat{B}'\Psi)^*\end{aligned}$$

so

$$(\Psi,[\hat{A}\hat{B}-\hat{B}\hat{A}]\Psi) = 2i\text{Im}\,(\hat{A}'\Psi,\hat{B}'\Psi)$$

In the last two steps we have made use of Eqs. (2-34a) and (2-13), respectively. We now take the modulus of both sides of this equation. Using the fact that $|i| = 1$, along with Eq. (2-18a), we obtain

$$|(\Psi,[\hat{A}\hat{B}-\hat{B}\hat{A}]\Psi)| = 2|\text{Im}\,(\hat{A}'\Psi,\hat{B}'\Psi)|$$

According to Eqs. (2-17), the magnitude of the imaginary part of a complex number is never greater than the modulus of the complex number, so we may write

$$|(\Psi,[\hat{A}\hat{B}-\hat{B}\hat{A}]\Psi)| \leqq 2|(\hat{A}'\Psi,\hat{B}'\Psi)|$$

†The expression for $\Delta\hat{A}$ in Lemma 3 allows a very simple proof of the fact that *the vanishing of* $\Delta\hat{A}_t$ *is a necessary and sufficient condition for* $\Psi_t(x)$ *to be coincident with an eigenvector of* $\hat{A}$. For, according to Lemma 3, $\Delta\hat{A}_t$ vanishes if and only if the vector $\hat{A}'\Psi_t(x) \equiv (\hat{A}-\langle\hat{A}\rangle_t)\Psi_t(x)$ has zero norm. But the only vector with zero norm is the null vector, $f(x) \equiv 0$. Therefore, $\Delta\hat{A}_t = 0$ if and only if $\hat{A}\Psi_t(x) \equiv \langle\hat{A}\rangle_t\Psi_t(x)$—i.e., if and only if $\Psi_t(x)$ is an eigenvector of $\hat{A}$. We had previously inferred the *sufficiency* condition only indirectly through Exercises 29 and 33.

We now write the Schwarz inequality, Eq. (2-34d), with $\psi_1(x) = \hat{A}'\Psi(x)$ and $\psi_2(x) = \hat{B}'\Psi(x)$:

$$|(\hat{A}'\Psi, \hat{B}'\Psi)| \leqq \sqrt{(\hat{A}'\Psi, \hat{A}'\Psi)} \cdot \sqrt{(\hat{B}'\Psi, \hat{B}'\Psi)}$$

Applying this to the previous inequality, we obtain

$$|(\Psi,[\hat{A}\hat{B} - \hat{B}\hat{A}]\Psi)| \leqq 2\sqrt{(\hat{A}'\Psi, \hat{A}'\Psi)} \cdot \sqrt{(\hat{B}'\Psi, \hat{B}'\Psi)}$$

or, by virtue of Lemma 3,

$$|(\Psi,[\hat{A}\hat{B} - \hat{B}\hat{A}]\Psi)| \leqq 2\Delta\hat{A} \cdot \Delta\hat{B}$$

This last equation is just Eq. (4-24), and so the proof is complete.

4-4 TIME EVOLUTION OF THE QUANTUM STATE

In our review of classical mechanics in Chapter 3 we considered at some length the problem of the *time evolution* of the classical state, but we did not deem it necessary to dwell very long on the rather obvious *definition* of the classical state. By contrast, our comparatively lengthy sojourn into quantum mechanics thus far has been almost exclusively concerned with describing precisely what is meant by the "state of a system" in quantum mechanics—a task which we have found to be anything but trivial. For, since quantum mechanics makes a radical, seemingly perverse distinction between the *state* of a system and the physical *observables*, it was necessary first to define each of these concepts separately, and then, using the quantum theory of measurement, to carefully delineate the subtle relationship between the two. Fortunately, however, once we arrive at a reasonably good understanding of these concepts, it is not too difficult to comprehend what the theory of quantum mechanics has to say about how a system behaves with time. As we shall see, in this respect quantum mechanics has much in common with the classical approach, in that it provides us with well-defined "equations of motion" for the state vector $\Psi_t(x)$, and also for the expectation value $\langle\hat{A}\rangle_t$ and the probability coefficients (α_k,Ψ_t) for a given observable $\mathcal{A}$. In this section we shall obtain these equations of motion and discuss some of their important implications.

In order that our development will be valid for *any* physical system with one degree of freedom, we shall continue to refer to observables "in general" by $\mathcal{A}, \mathcal{B}$, etc. In Sec. 4-5 we shall exhibit the specific forms of the relevant observable operators for the particular case of a mass m on the x-axis (this will constitute our Postulate 6), and we shall then apply our general theory to that particular problem in some detail.

4-4a Energy and the Hamiltonian Operator The Time Evolution Equation for $\Psi_t(x)$

In our discussion of the time-evolution problem in classical mechanics, we found that there was a very intimate connection between *energy* and *time*. Thus, while on the one hand the observable "energy" was unique in that it maintained a constant value in time [see Exercise 22], it turned out that it was the energy written as the Hamiltonian function $H(x,p)$ that actually governed the time development of the system [see Eqs. (3-10)]. Keeping this in mind, we begin our discussion of the time-evolution problem in quantum mechanics by presenting the postulate which tells us exactly how the state vector of a system changes with time.

Postulate 5. For every physical system there exists a linear, Hermitian operator $\hat{\mathrm{H}}$, called the *Hamiltonian operator*, which has the following properties:

(a) The Hamiltonian operator $\hat{\mathrm{H}}$ is the observable operator corresponding to the total energy of the system. Hence, $\hat{\mathrm{H}}$ possesses a complete, orthonormal set of eigenvectors $\{\eta_k(x)\}$ and a corresponding set of real eigenvalues $\{E_k\}$,

$$\hat{\mathrm{H}}\eta_k(x) = E_k\eta_k(x) \qquad k = 1, 2, \ldots \tag{4-25}$$

where the numbers $\{E_k\}$ are the allowed values of the total energy of the system.

(b) The Hamiltonian operator $\hat{\mathrm{H}}$ determines the time evolution of the state vector of the system, $\Psi_t(x) \equiv \Psi(x,t)$, through the differential equation

$$i\hbar \frac{\partial}{\partial t} \Psi(x,t) = \hat{\mathrm{H}}\Psi(x,t) \tag{4-26}$$

provided the system is not disturbed. The constant $\hbar$ is called "h-bar," and has the value

$$\hbar \equiv h/2\pi = 1.054 \times 10^{-34} \text{ joule} \cdot \text{sec} \tag{4-27}$$

According to this postulate, if $\Psi_0(x)$ is the state vector of the system at time $t = 0$, then (provided the system is not intruded upon by some external agency) as t assumes successive values $t_1, t_2, \ldots,$ the state vector successively coincides with the $\mathcal{H}$-vectors $\Psi_{t_1}(x) \equiv \Psi(x,t_1)$, $\Psi_{t_2}(x) \equiv \Psi(x,t_2), \ldots,$ where $\Psi(x,t)$ is that solution of Eq. (4-26) which satisfies the "initial condition" $\Psi(x,0) \equiv \Psi_0(x)$. Consequently, the state vector of the system evolves with time in a completely deterministic way, just as $x(t)$ and $p(t)$ do in classical mechanics. This is true so long as the system is not disturbed, and in this connection we must emphasize that, according to Postulate 4, *the measurement process represents such a disturbance* which alters the otherwise orderly time development of the state vector.

We may now trace the following interesting parallelism between classical mechanics and quantum mechanics: In {classical | quantum} mechanics, once the Hamiltonian {function, $H(x,p)$ | operator, $\hat{H}$} is specified for a given system, then the undisturbed time evolution of the state of the system {$[x(t),p(t)]$ | $\Psi(x,t)$} can be uniquely determined by solving the time-evolution equation {Eqs. (3-10)| Eq. (4-26)}, subject to the specified initial condition {$[x(0),p(0)] = [x_0, p_0]$ | $\Psi(x,0) = \Psi_0(x)$}. Pursuing this parallelism a bit further, we recall that the postulates of classical mechanics did *not* specify the exact *form* which the Hamiltonian function assumes for a given physical system; for example, Newton's second law does not tell us that the Hamiltonian function appropriate to a mass m attached to a spring is $H(x,p) = p^2/2m + kx^2/2$. Similarly, Postulate 5 does *not* tell us what to write down for the Hamiltonian operator $\hat{H}$ for a given physical system. Indeed, it is the task of the physicist as a "clever observer of Nature" to *discover* or *invent* an appropriate Hamiltonian {function, $H(x,p)$| operator, $\hat{H}$} for a given {classical | quantum} system.

We see then that, given the explicit form of the Hamiltonian operator, we can in principle calculate the state vector of the system at time t *if* we know the state vector at time 0. An obvious question at this point is, how can we ever know what the state vector is at $t = 0$? Unless we are just "given" $\Psi_0(x)$ outright, the only way we can know what it is for sure is to *make a measurement* of some observable $\mathcal{A}$ at time $t = 0$: By Postulate 4, this measurement will force the system into one of the eigenvectors of $\hat{A}$, and merely by taking cognizance of which eigenvalue was obtained in the measurement, we will know *which* eigenvector the system was forced into. Thus, if a measurement of $\mathcal{A}$ at time 0 yields the result A_k, then immediately after the measurement the state vector will begin to evolve from the vector $\Psi_0(x) = \alpha_k(x)$ in the manner prescribed by Eq. (4-26). This orderly time evolution will continue until such time as another measurement is performed on the system, at which time the foregoing process will be repeated. It will be observed that this method of "preparing" the state $\Psi_0(x)$ depends crucially upon the fact that the measurement operation tells us what the state vector is immediately *after* a measurement, rather than immediately *before*.

One of the requirements of Postulate 1 was that the state vector of the system always have unit norm: $(\Psi_t, \Psi_t) = 1$. It is reasonable to ask whether or not the time behavior of $\Psi_t(x)$, as specified by

Eq. (4-26), is such that, if $(\Psi_0,\Psi_0) = 1$, then it will be true that $(\Psi_t,\Psi_t) = 1$ for all $t > 0$. To show that this is indeed the case, we shall prove that the time derivative of the norm of $\Psi_t(x)$ vanishes identically. We have

$$\frac{d}{dt}(\Psi_t,\Psi_t) = \left(\frac{\partial\Psi_t}{\partial t},\Psi_t\right) + \left(\Psi_t,\frac{\partial\Psi_t}{\partial t}\right) \tag{4-28}$$

Exercise 40. By writing out the inner product (Ψ_t,Ψ_t) in its integral form, $\int\Psi^*(x,t)\Psi(x,t)dx$, and using the fact that the integration variable x is independent of t, prove Eq. (4-28).

Now according to Eq. (4-26),

$$\frac{\partial\Psi_t}{\partial t} = \frac{1}{i\hbar}\hat{H}\Psi_t = -\frac{i}{\hbar}\hat{H}\Psi_t$$

where we have used the fact that, since $i^2 = -1$, then $1/i = -i$. Inserting this into the right-hand side of Eq. (4-28), we obtain

$$\begin{aligned}\frac{d}{dt}(\Psi_t,\Psi_t) &= \left(-\frac{i}{\hbar}\hat{H}\Psi_t,\Psi_t\right) + \left(\Psi_t,-\frac{i}{\hbar}\hat{H}\Psi_t\right)\\ &= \left[-\frac{i}{\hbar}\right]^*(\hat{H}\Psi_t,\Psi_t) + \left[-\frac{i}{\hbar}\right](\Psi_t,\hat{H}\Psi_t)\\ &= \left[\frac{i}{\hbar}\right][(\hat{H}\Psi_t,\Psi_t) - (\Psi_t,\hat{H}\Psi_t)]\end{aligned}$$

Since $\hat{H}$ is an Hermitian operator, then for any state $\Psi_t(x)$, we have $(\hat{H}\Psi_t,\Psi_t) = (\Psi_t,\hat{H}\Psi_t)$; therefore, we conclude that

$$\frac{d}{dt}(\Psi_t,\Psi_t) = 0 \tag{4-29}$$

This proves that the time evolution of the state vector, as dictated by Postulate 5, is consistent with the requirement of Postulate 1 that the state vector always have unit norm.

Although the time evolution of the state vector is completely specified by the differential equation (4-26), there is another way of describing this time evolution which has a formal elegance, and occasional usefulness, that Eq. (4-26) lacks. According to Postulate 5, if we place the system in some particular state $\Psi_0(x)$ at time 0, then the passage of a time t "transforms" this state vector into some new vector $\Psi_t(x)$, which can be found by solving Eq. (4-26). However, we recall from our discussion in Sec. 2-4 that the transformation of a given $\mathcal{H}$-vector into another $\mathcal{H}$-vector may generally be regarded as the result of an *operator* acting on the given vector. In this spirit, it is tempting to write

$$\Psi_t(x) = \hat{U}(t)\Psi_0(x) \tag{4-30}$$

where $\hat{U}(t)$ is some operator that transforms the state vector at time 0 into the state vector at time t. Now, we know from Eq. (4-26) that the time evolution of the state vector depends crucially upon the system's Hamiltonian operator $\hat{H}$, so it is reasonable to expect that $\hat{U}(t)$ will be some function of $\hat{H}$ as well as t. It is also possible that $\hat{U}(t)$ might depend upon $\Psi_0(x)$ as well, but if this were the case then Eq. (4-30) would not be very interesting or useful. However, by using the definition of $\hat{U}(t)$ in Eq. (4-30) together with the time-evolution equation (4-26), we can show that the operator $\hat{U}(t)$ is a well-defined function of $\hat{H}$ and t alone, and is independent of the initial state $\Psi_0(x)$. We call $\hat{U}(t)$ the *time-evolution operator* of the system.

To deduce the form of the time-evolution operator, we first substitute Eq. (4-30) into Eq. (4-26):

$$i\hbar \frac{\partial \hat{U}(t)}{\partial t} \Psi_0(x) = \hat{H}\hat{U}(t)\Psi_0(x)$$

This equation says that the operator $i\hbar(\partial \hat{U}/\partial t)$ acting on the $\mathcal{H}$-vector $\Psi_0(x)$ must produce the same $\mathcal{H}$-vector as does the operator $\hat{H}\hat{U}$ acting on $\Psi_0(x)$. Since we want this to be true for *any* choice of $\Psi_0(x)$, then we must have

$$i\hbar \frac{\partial \hat{U}(t)}{\partial t} = \hat{H}\hat{U}(t) \tag{4-31}$$

This is a differential equation for the operator $\hat{U}(t)$. The time-evolution operator must satisfy this differential equation for all $\hat{H}$ and t, and it must *also* satisfy the "initial condition"

$$\hat{U}(0) = 1 \tag{4-32}$$

Exercise 41. Derive Eq. (4-32) from the definition of $\hat{U}(t)$ in Eq. (4-30), and explain its meaning.

We shall now demonstrate that the required form for the operator $\hat{U}(t)$ is

$$\hat{U}(t) = 1 + \sum_{n=1}^{\infty} \frac{\left(-\frac{i}{\hbar}\hat{H}t\right)^n}{n!} \equiv e^{-i\hat{H}t/\hbar} \tag{4-33}$$

This equation asserts that $\hat{U}(t)$ is a certain power series in the operator $\hat{H}$ (the power series has a well-defined meaning by virtue of Eqs. (2-42)), and we have chosen to denote this power series by the *symbol* $\exp(-i\hat{H}t/\hbar)$. In adopting this symbolic representation for the power series in Eq. (4-33), we are merely following the procedure described in Sec. 4-2 for defining functions of operators [see Eqs. (4-5)]; for, it will be recalled that the function $f(x) = e^x$ has the Taylor series expansion

$$e^x = 1 + \sum_{n=1}^{\infty} \frac{x^n}{n!}.$$

and we have merely replaced x by $(-i\hat{H}t/\hbar)$ in obtaining the right-hand side of Eq. (4-33).

It is easy to see that the power series in Eq. (4-33) satisfies the initial condition Eq. (4-32), since all the terms under the summation sign vanish for $t = 0$. To see that this power series also satisfies the differential equation (4-31), we calculate its time derivative. Remembering that the operator $\hat{H}$ is linear, so that $(\hat{H}t)^n = \hat{H}^n t^n$, and also that $\hat{H}$ is independent of t, we have from Eq. (4-33)

$$\frac{\partial \hat{U}(t)}{\partial t} = 0 + \sum_{n=1}^{\infty} \frac{\left(-\frac{i}{\hbar}\hat{H}\right)^n (nt^{n-1})}{n!} = \left(-\frac{i}{\hbar}\hat{H}\right) \sum_{n=1}^{\infty} \frac{\left(-\frac{i}{\hbar}\hat{H}t\right)^{n-1}}{(n-1)!}$$

$$= \left(-\frac{i}{\hbar}\hat{H}\right)\left[1 + \sum_{n=1}^{\infty} \frac{\left(-\frac{i}{\hbar}\hat{H}t\right)^n}{n!}\right] = \frac{1}{i\hbar}\hat{H}\hat{U}(t)$$

which is identical to Eq. (4-31). Consequently, the operator $U(t)$ in Eq. (4-33) is indeed the time-evolution operator of the system; it satisfies Eqs. (4-31) and (4-32), and therefore it satisfies Eq. (4-30) for all $t \geqq 0$ and all choices of the initial-state vector $\Psi_0(x)$.

Exercise 42. Show that if one ignores the operator character of $\hat{H}$ and $\hat{U}(t)$, then the "symbol" $\exp(-i\hat{H}t/\hbar)$ satisfies Eqs. (4-31) and (4-32) in a purely formal sense.

Since $\hat{U}(t)$ in Eq. (4-33) is a linear combination of products of the linear operator $\hat{H}$, it follows that $\hat{U}(t)$ is a *linear* operator. However, the coefficients in this linear combination are obviously not real, so it does not follow that $\hat{U}(t)$ is an Hermitian operator [see Exercise 16]. In fact, it is easy to see that $\hat{U}(t)$ is *not* Hermitian: We recall from Exercise 26 that the operator $f(\hat{A})$ has eigenvectors $\{\alpha_n(x)\}$ and eigenvalues $\{f(A_n)\}$, where $\{\alpha_n(x)\}$ and $\{A_n\}$ are the eigenvectors and eigenvalues of $\hat{A}$. Therefore, the operator $\hat{U}(t)$ has eigenvectors $\{\eta_n(x)\}$ and eigenvalues $\{\exp(-iE_n t/\hbar)\}$. In particular, since the eigenvalues of U(t) are not pure real [see Eq. (2-20a)], then $\hat{U}(t)$ cannot be Hermitian [see the discussion preceding Exercise 18]. Therefore, $\hat{U}(t)$ is *not* to be regarded as an *observable* operator; its character and function are altogether different from the operators which we have discussed so far.

The time-evolution operator plays an important role in more advanced treatments of the time behavior of systems; however, in our subsequent work in this book we shall use the time-evolution *equation*, Eq. (4-26), rather that the time-evolution *operator*, Eq. (4-33). We have introduced the time-evolution operator primarily because the picture conveyed by Eq. (4-30), of $\Psi_0(x)$ being "carried into" $\Psi_t(x)$ by a linear operator which depends only on $\hat{H}$ and t, is conceptually very important, and significantly enhances our appreciation of the time-evolution process in quantum mechanics.

It was pointed out earlier that the general observable operator $\hat{A}$, along with its eigenbasis $\{\alpha_n(x)\}$ and eigenvalues $\{A_n\}$, do *not* depend upon time. However, the time-dependence of $\Psi_t(x)$ prescribed by Postulate 5 clearly implies a time-dependence for the expectation value $\langle\hat{A}\rangle_t = (\Psi_t,\hat{A}\Psi_t)$, and the probability coefficients (α_i,Ψ_t). In the next two sections we shall derive from Eq. (4-26) the time-evolution equations for these two quantities.

4-4b The Time Evolution Equation for $\langle\hat{A}\rangle_t$ The Time-Energy Uncertainty Relation

We found in Sec. 4-3a that the results of many repeated measurements of an observable $\mathcal{A}$ on a state $\Psi_t(x)$ can often be adequately described by the expectation value $\langle\hat{A}\rangle_t$ and the uncertainty $\Delta\hat{A}_t$ [see Fig. 3]. The behavior of these quantities with time is therefore of some interest. We shall not examine here the time evolution of $\Delta\hat{A}_t$, but we note that, by virtue of Eq. (4-14), the time behavior of $\Delta\hat{A}_t$ can in principle be determined if one knows how to determine the time behavior of expectation values. It is this latter problem that we shall examine in this section.

To investigate the time dependence of $\langle\hat{A}\rangle_t$, let us simply calculate its time derivative. This calculation is formally very similar to the one carried out in the last section in proving Eq. (4-29). Using the fact that the inner-product integration variable x and the observable operator $\hat{A}$ are both independent of time, we have in analogy with Eq. (4-28),

$$\frac{d}{dt}\langle\hat{A}\rangle_t = \frac{d}{dt}(\Psi_t,\hat{A}\Psi_t) = \left(\frac{\partial\Psi_t}{\partial t},\hat{A}\Psi_t\right) + \left(\Psi_t,\hat{A}\frac{\partial\Psi_t}{\partial t}\right)$$

Then, inserting the expression for $\partial\Psi_t/\partial t$ given in Postulate 5,

$$\frac{d}{dt}\langle\hat{A}\rangle_t = \left(-\frac{i}{\hbar}\hat{H}\Psi_t,\hat{A}\Psi_t\right) + \left(\Psi_t,\hat{A}\left[-\frac{i}{\hbar}\hat{H}\Psi_t\right]\right)$$

$$= \left[-\frac{i}{\hbar}\right]^*(\hat{H}\Psi_t,\hat{A}\Psi_t) + \left[-\frac{i}{\hbar}\right](\Psi_t,\hat{A}\hat{H}\Psi_t)$$

$$= \left[\frac{i}{\hbar}\right][(\hat{H}\Psi_t,\hat{A}\Psi_t) - (\Psi_t,\hat{A}\hat{H}\Psi_t)]$$

Finally, invoking the Hermiticity of $\hat{H}$, we obtain

$$\frac{d}{dt}\langle\hat{A}\rangle_t = \left[\frac{i}{\hbar}\right][(\Psi_t,\hat{H}\hat{A}\Psi_t) - (\Psi_t,\hat{A}\hat{H}\Psi_t)]$$

or

$$\frac{d}{dt}\langle\hat{A}\rangle_t = \frac{i}{\hbar}(\Psi_t,[\hat{H}\hat{A} - \hat{A}\hat{H}]\Psi_t) \tag{4-34}$$

This is the fundamental time-evolution equation for the expectation value of an observable $\mathcal{A}$. It evidently gives the instantaneous time-rate-of-change of $\langle\hat{A}\rangle_t$ in terms of the instantaneous state vector $\Psi_t(x)$ and the operator $(\hat{H}\hat{A} - \hat{A}\hat{H})$.

Exercise 43. Using Eq. (4-34), prove that if $\hat{A}$ commutes with the Hamiltonian operator of the system, then both the expectation value $\langle\hat{A}\rangle_t$ and the uncertainty $\Delta\hat{A}_t$ are constant in time. [*Hint*: To prove that $\Delta\hat{A}_t$ is constant, show that both $\langle\hat{A}^2\rangle_t$ and $\langle\hat{A}\rangle_t^2$ are constants.]

The foregoing exercise provides us with a rule for determining the "constants of the motion": If $\hat{A}$ commutes with $\hat{H}$, or equivalently, if $\mathcal{A}$ is compatible with the total energy of the system, then $\mathcal{A}$ is a constant of the motion in the sense that $\langle\hat{A}\rangle_t$ and $\Delta\hat{A}_t$ are independent of time. In particular, since $\hat{H}$ certainly commutes with itself, then $\langle\hat{H}\rangle_t$ and $\Delta\hat{H}_t$ are *always* constant in time. This is analogous to the result in classical mechanics that the energy is a constant of the motion [see Exercise 22]. We shall discuss the concept of a "constant of the motion" in greater detail in the next section.

One aspect of the relationship between time and energy which has no direct analogue in classical mechanics is the so-called "time-energy uncertainty principle." We define for a given observable $\mathcal{A}$ its *evolution time* $T_{\mathcal{A}}$ by

$$T_{\mathcal{A}} \equiv \Delta\hat{A}_t \Big/ \left|\frac{d\langle\hat{A}\rangle_t}{dt}\right| \tag{4-35}$$

To understand the physical significance of $T_\mathcal{A}$, let us first suppose that $d\langle\hat{A}\rangle_t/dt$ is constant in time. Then, in a given time interval Δt, the expectation value of $\mathcal{A}$ would change by an amount $|d\langle\hat{A}\rangle_t/dt|\cdot\Delta t$; in particular, we see from Eq. (4-35) that in a time interval equal to $T_\mathcal{A}$, $\langle\hat{A}\rangle_t$ would change by an amount equal to $\Delta\hat{A}_t$. In general, even if $d\langle\hat{A}\rangle_t/dt$ is *not* a constant, it is obvious that $T_\mathcal{A}$, as defined by Eq. (4-35), provides a very reasonable estimate of the amount of of time which must elapse before the expectation value of $\mathcal{A}$ changes by an amount equal to the uncertainty in $\mathcal{A}$. In other words, $T_\mathcal{A}$ is the time which must elapse before the *average* of the values measured for $\mathcal{A}$ in a series of repeated measurements, $\langle\hat{A}\rangle_t$, changes or evolves enough to be *noticeable* over the *intrinsic spread* in these values, $\Delta\hat{A}_t$. It is in this sense that $T_\mathcal{A}$ specifies the "evolution time" of $\mathcal{A}$.

Now, by combining the time-evolution equation for $\langle\hat{A}\rangle_t$, Eq. (4-34), with the generalized uncertainty relation, Eq. (4-24), we can easily derive the following inequality:

$$T_\mathcal{A}\cdot\Delta\hat{H}\geqq\frac{\hbar}{2}\qquad(4\text{-}36)$$

Exercise 44. Derive Eq. (4-36). [*Hint;* First obtain an expression for $|d\langle\hat{A}\rangle_t/dt|$ from Eq. (4-34); then make use of Eqs. (4-24) and (4-35).]

The above inequality is called the *time-energy uncertainty relation*. It states that the more precisely the energy of a system is defined (i.e., the smaller $\Delta\hat{H}$ is), then the more slowly will *any* observable $\mathcal{A}$ change "noticeably" with time (i.e., the larger any $T_\mathcal{A}$ must be); conversely, if *any* observable plainly exhibits a rapid variation with time (i.e., if any $T_\mathcal{A}$ is small), then the system *cannot* have a well-defined energy (i.e., $\Delta\hat{H}$ must be large). These predictions of Eq. (4-36), although obviously "nonclassical" in character, have been amply confirmed in the laboratory by spectroscopic studies comparing the "widths" of excited atomic energy levels with the corresponding "lifetimes" of these levels: it is found that narrow, sharp energy levels have long lifetimes, while broad, diffuse energy levels have short lifetimes.

4-4c The Time Evolution Equation for (α_n,Ψ_t)
Constants of the Motion and Stationary States

There are two reasons why it is useful to know the time-dependence of the quantities

$$a_n(t)\equiv(\alpha_n,\Psi_t)\qquad n=1,2,\ldots\qquad(4\text{-}37)$$

First, the square moduli of these quantities determine the shape of the distribution curve for $\mathcal{A}$ [see Fig. 3], and it is clearly of interest to know how this curve changes with time. Second, since the state vector of the system can always be written in the form

$$\Psi_t(x) = \sum_{n=1}^{\infty} (\alpha_n, \Psi_t)\alpha_n(x) \equiv \sum_{n=1}^{\infty} a_n(t)\alpha_n(x) \qquad (4\text{-}38)$$

then if a simple expression for $a_n(t)$ can be found, we will have an explicit representation for the time-varying state vector.

To see what we can discover about the time-dependence of $a_n(t)$, let us calculate its time derivative. Since $\alpha_n(x)$ is independent of t, we have

$$\frac{d}{dt}a_n(t) \equiv \frac{d}{dt}(\alpha_n, \Psi_t) = \left(\alpha_n, \frac{\partial \Psi_t}{\partial t}\right)$$

Inserting the expression for $\partial \Psi_t / \partial t$ given in Postulate 5,

$$\frac{d}{dt}a_n(t) = \left(\alpha_n, -\frac{i}{\hbar}\hat{H}\Psi_t\right) = -\frac{i}{\hbar}(\alpha_n, \hat{H}\Psi_t)$$

We now replace $\Psi_t(x)$ by its expansion in Eq. (4-38), and make use of the linearity of $\hat{H}$:

$$\frac{d}{dt}a_n(t) = -\frac{i}{\hbar}\left(\alpha_n, \hat{H}\sum_{m=1}^{\infty} a_m(t)\alpha_m\right) = -\frac{i}{\hbar}\left(\alpha_n, \sum_{m=1}^{\infty} a_m(t)\hat{H}\alpha_m\right)$$

Finally, using Eqs. (2-34b) and (2-34c), we obtain the result

$$\frac{d}{dt}a_n(t) = -\frac{i}{\hbar}\sum_{m=1}^{\infty} a_m(t)\,(\alpha_n, \hat{H}\alpha_m) \qquad n = 1, 2, \ldots \qquad (4\text{-}39)$$

For an arbitrary eigenbasis $\{\alpha_i(x)\}$, this is about as far as we can go. Recognizing that the quantities $(\alpha_n, \mathrm{H}\alpha_m)$ in Eqs. (4-39) are ordinary complex numbers, we see that Eqs. (4-39) express the time derivative of *each* $a_n(t)$ as a linear combination of *all* the $a_i(t)$. In principle, one could solve this infinite set of "coupled, linear differential equations," and so obtain an explicit expression for each $a_n(t)$; in practice, however, this is usually far too difficult to do for an arbitrary eigenbasis $\{\alpha_i(x)\}$.

There is, however, one important case for which Eqs. (4-39) can be solved fairly easily. Suppose the eigenbasis $\{\alpha_i(x)\}$ coincides with the *energy* eigenbasis $\{\eta_i(x)\}$; this will be true if the observable operator $\hat{A}$ under consideration coincides with $\hat{H}$, or more generally, by the Compatibility Theorem, if $\hat{A}$ *commutes* with $\hat{H}$. In this case, it is easy to show that Eqs. (4-39) "uncouple" and take the simple forms

$$\frac{d}{dt}(\eta_n, \Psi_t) = -\frac{i}{\hbar}E_n(\eta_n, \Psi_t) \qquad n = 1,2, \ldots \tag{4-40}$$

Exercise 45. Show that Eqs. (4-39) reduce to Eqs. (4-40) when the eigenvectors $\{\alpha_i(x)\}$ coincide with the energy eigenvectors $\{\eta_i(x)\}$ as defined in Eq. (4-25).

Eqs. (4-40) can be immediately integrated: Writing it as

$$\frac{d(\eta_n, \Psi_t)}{(\eta_n, \Psi_t)} = -\frac{iE_n}{\hbar}dt$$

then an elementary integration yields

$$\log(\eta_n, \Psi_t) = -\frac{iE_n}{\hbar}t + \log C$$

where we have written the integrating constant as log C. Using the properties of the exponential, we can write this last equation as

$$(\eta_n, \Psi_t) = C\mathrm{e}^{-iE_n t/\hbar}$$

where the complex exponential has been defined and discussed in Exercise 7. Finally since $\mathrm{e}^0 = 1$, we see that we must set $C = (\eta_n, \Psi_0)$; consequently, we conclude that †

$$(\eta_n, \Psi_t) = (\eta_n, \Psi_0)\mathrm{e}^{-iE_n t/\hbar} \qquad n = 1,2, \ldots \tag{4-41}$$

Thus while it is not possible to say very much about the time dependence of (α_n, Ψ_t) for an *arbitrary* eigenbasis $\{\alpha_i(x)\}$, Eqs. (4-41) give the *explicit* time dependence for the case in which the eigenbasis is the *energy* eigenbasis. For the remainder of this section, we shall examine some of the consequences of Eqs. (4-41).

Let us first discuss Eqs. (4-41) from the viewpoint that $|(\eta_n, \Psi_t)|^2$ represents a *probability*. We recall from Exercise 43 that any observable $\mathfrak{A}$ whose corresponding operator $\hat{A}$ commutes with $\hat{H}$ is a sort of "constant of the motion" in the sense that $\langle\hat{A}\rangle_t$ and $\Delta\hat{A}_t$ do not change with time. Now if $\hat{A}$ commutes with $\hat{H}$, then according to the Compatibility Theorem, $\hat{A}$ and $\hat{H}$ must share the same

†Our derivation of Eq. (4-41) may leave the reader a bit uneasy, since it involved the complex logarithm, a concept which we have not discussed. However, the reader can easily verify that the formula for (η_n, Ψ_t) given in Eqs. (4-41) does indeed satisfy the differential equations (4-40), simply by invoking Eq. (2-20e). Since Eqs. (4-41) satisfy *both* the differential equations and the initial conditions, it is therefore *the* solution.

eigenbasis; thus, by making at most a rearrangement of indices, we can put

$$\alpha_n(x) = \eta_n(x) \qquad n = 1, 2, \ldots$$

This being the case, Eqs. (4-41) imply that

$$(\alpha_n, \Psi_t) = (\eta_n, \Psi_t) = (\eta_n, \Psi_0)\, e^{-iE_n t/\hbar} = (\alpha_n, \Psi_0)\, e^{-iE_n t/\hbar}$$

Therefore, the probability of measuring for $\mathcal{A}$ the eigenvalue A_n at time t is

$$|(\alpha_n, \Psi_t)|^2 = |(\alpha_n, \Psi_0)|^2 |e^{-iE_n t/\hbar}|^2 = |(\alpha_n, \Psi_0)|^2$$

where we have made use of Eqs. (2-18a) and (2-20d). We have thus shown that, if $\hat{A}$ commutes with $\hat{H}$, then the probability of measuring for $\mathcal{A}$ the eigenvalue A_n is the same at time t as at time 0. Since the quantities $|(\alpha_n, \Psi_t)|^2$ determine the distribution curve for $\mathcal{A}$ at time t [see Fig. 3], then we see that, if $\hat{A}$ commutes with $\hat{H}$, not only are the "center" and "width" of the distribution curve constant in time [see Exercise 43], but indeed the *entire curve* does not change with time. We are therefore quite justified in calling $\mathcal{A}$ a *constant of the motion* whenever $\hat{A}$ commutes with $\hat{H}$.

Let us next examine the consequences of Eqs. (4-41) with regard to the *representation of the time-varying state vector.* Since we can always expand $\Psi_t(x)$ in the eigenbasis of $\hat{H}$,

$$\Psi_t(x) = \sum_{n=1}^{\infty} (\eta_n, \Psi_t)\eta_n(x)$$

then substituting Eqs. (4-41) yields at once

$$\Psi_t(x) = \sum_{n=1}^{\infty} (\eta_n, \Psi_0) e^{-iE_n t/\hbar} \eta_n(x) \tag{4-42}$$

Inasmuch as the complex numbers (η_n, Ψ_0) do not depend upon time, Eq. (4-42) shows *explicitly* the time dependence of the state vector $\Psi_t(x)$, and therefore represents a *general solution to the fundamental time-evolution equation* (4-26). We note in particular that, since $e^0 = 1$, then for $t = 0$ Eq. (4-42) reduces to

$$\Psi_0(x) = \sum_{n=1}^{\infty} (\eta_n, \Psi_0)\eta_n(x) \tag{4-43}$$

which of course is an identity for *any* $\mathcal{H}$-vector $\Psi_0(x)$ [see Eq. (4-2b)]. Indeed, an easy way to remember Eq. (4-42) is to first write down Eq. (4-43), and then simply insert the factor $\exp(-iE_n t/\hbar)$ in the *n*th term of the sum.† We have in Eq. (4-42) yet another instance of the intimate connection between time and energy: if we can find the energy eigenvectors $\{\eta_i(x)\}$ and eigenvalues $\{E_i\}$—i.e., if we can solve the energy eigenvalue equation (4-25)—then we can write down at once a complete solution to the time-evolution equation for the state vector, Eq. (4-26).

Exercise 46. By directly substituting Eq. (4-42) into Eq. (4-26), show that this expression for $\Psi_t(x)$ does indeed satisfy the time-evolution equation of Postulate 5. [*Hint*: Make use of Eqs. (2-20e) and (4-25), and remember that neither $\partial/\partial t$ nor $\hat{H}$ has any effect upon the complex constants (η_n, Ψ_0).]

It is interesting to consider the special case in which the initial state vector, $\Psi_0(x)$, coincides with one of the energy eigenvectors, say $\eta_k(x)$. In this case, the coefficients in Eq. (4-42) are

$$(\eta_n, \Psi_0) = (\eta_n, \eta_k) = \delta_{nk}$$

and Eq. (4-42) reduces to

$$\Psi_t(x) = e^{-iE_k t/\hbar}\, \eta_k(x)$$

This says that $\Psi_t(x)$ differs from $\Psi_0(x) = \eta_k(x)$ only by a scalar factor whose square modulus is *unity* [see Eq. (2-20d)]:

$$|e^{-iE_k t/\hbar}|^2 = 1$$

Consequently, by Postulate 1, the state of the system at time t is *physically the same* as the state at time 0. In such a case, it is clear that *all* observables will behave like "constants of the motion."

Exercise 47.

(a) Prove directly that if $\Psi_t(x) = e^{-iE_k t/\hbar}\eta_k(x)$, then for *any* observable $\mathcal{A}$, the expectation value $\langle \hat{A} \rangle_t$, the uncertainty $\Delta \hat{A}_t$, and the probabilities $|(\alpha_n, \Psi_t)|^2$ are all constant in time—regardless of whether or not $\hat{A}$ commutes with $\hat{H}$.

†Equation (4-42) can also be *derived* from Eq. (4-43) by using the time evolution operator $\hat{U}(t) = \exp(-i\hat{H}t/\hbar)$, which was discussed in Sec. 4-4a. We recall that this operator is a *linear* operator, and has eigenvectors $\{\eta_n(x)\}$ and eigenvalues $\{\exp(-iE_n t/\hbar)\}$. Therefore,

$$\Psi_t(x) = \hat{U}(t)\Psi_0(x) = \hat{U}(t) \sum_{n=1}^{\infty} (\eta_n, \Psi_0)\eta_n(x)$$

$$= \sum_{n=1}^{\infty} (\eta_n, \Psi_0)\hat{U}(t)\eta_n(x) = \sum_{n=1}^{\infty} (\eta_n, \Psi_0)\, e^{-iE_n t/\hbar}\, \eta_n(x)$$

which is evidently Eq. (4-42).

(b) Show that in this case we will have at any time t, $\Delta\hat{H} = 0$ and $T_a = \infty$. Discuss this result in the light of the Time-Energy Uncertainty Relation, Eq. (4-36). [*Hint*: Use Exercise 33 and the definition in Eq. (4-35).]

The foregoing considerations motivate us to define, for a given system with a given Hamiltonian, the set of *time-dependent* $\mathcal{H}$-vectors $\{\Psi_t^{(n)}(x)\}$:

$$\Psi_t^{(n)}(x) \equiv \Psi^{(n)}(x,t) \equiv e^{-iE_n t/\hbar}\,\eta_n(x) \qquad n = 1, 2, \ldots \quad (4\text{-}44)$$

According to the above arguments, if $\Psi_0(x)$ coincides with $\eta_n(x)$, then $\Psi_t(x)$ will be given by $\Psi_t^{(n)}(x)$. Moreover, by Exercise 47, if the state vector of the system coincides with $\Psi_t^{(n)}(x)$, then *all* observables behave like constants of the motion. For this reason, the vectors $\{\Psi_t^{(n)}(x)\}$ are called the *stationary states* of the system. In terms of these stationary states, Eq. (4-42) can be written

$$\Psi_t(x) = \sum_{n=1}^{\infty} (\Psi_0^{(n)}, \Psi_0)\,\Psi_t^{(n)}(x) \qquad (4\text{-}45)$$

Exercise 48. Prove that Eq. (4-45) follows from Eqs. (4-42) and (4-44).

Thus the stationary states of a system are important because, given any initial state vector $\Psi_0(x)$, the state vector at any subsequent time t can be written as a linear combination of the stationary states. The coefficients in this linear combination are the *time-independent* complex numbers $(\Psi_0^{(n)}, \Psi_0)$, which can evidently be evaluated if the stationary states and the initial state are known as functions of x.

We see then that $\Psi_t(x)$ can in principle always be found if the stationary states can be determined. According to Eq. (4-42), the determination of the stationary states is tantamount to finding the eigenvectors and eigenvalues of the energy operator, $\hat{H}$. For this reason, most of the actual "problem solving" in quantum mechanics is concerned with solving the energy eigenvalue equation (4-25) for various Hamiltonian operators—an enterprise which is usually very difficult from a mathematical point of view.

4-4d "Determinism" in Quantum Mechanics

We have now essentially completed the task of erecting the main conceptual framework of quantum mechanics. In order to *apply* the theory to a particular situation, it remains only to specify the *exact forms* of the relevant observable operators, and then to find their eigenvectors and eigenvalues. In the next section we shall discuss how this is done for one particularly important class of physical systems. But before doing so, it seems appropriate now to reconsider, from the viewpoint of quantum mechanics, our brief remarks in Sec. 3-3 on the matter of "determinism."

By "determinism" we mean here the general possibility of predicting exactly how the state of a system will *change* in any given circumstance. In classical mechanics, the change in the state of any system with time is in principle completely predictable (barring, of course, any unwarranted disturbance of the system by some external agent), and on this basis we concluded that classical mechanics implies a "deterministic" universe.

With regard to quantum mechanics, the situation from one point of view is very similar: In our discussion of Postulate 5 we saw that the state vector of a quantum system evolves with time in a completely predictable manner, and in this sense it may be said that quantum mechanics, like classical mechanics, is a "deterministic theory." However, in quantum mechanics the state of a system not only changes with the passage of time, but it also changes as a result of being measured. In our discussion of Postulates 3 and 4 we found that the change induced in the state vector by a measurement is *in principle* neither controllable nor predictable. That is, if we decide to measure $\mathcal{A}$ on a given state, it is usually not possible to know precisely which eigenvector of $\hat{A}$ the state will be forced into by the measurement. From this point of view, it is evident that quantum mechanics is *not* a completely deterministic theory.

If we choose to regard the entire universe as a single system, governed by one super-Hamiltonian operator, then since there is nothing "external" which can make a measurement on this system, we may justifiably assert that the state of the whole universe evolves with time in a completely deterministic way. If, however, we wish to consider only a *portion* of the universe as our system, omitting for example ourselves and our measuring apparatus, then we must evidently contend with a certain amount of indeterministic behavior every time we make a measurement upon the system.

The general problem of assessing the impact of quantum mechanics upon the concepts of determinism and causality is obviously a very intriguing and many-faceted one, and is as much in the domain of philosophy as physics. Without demeaning the importance of this complex problem, we shall not try to discuss it any further here. We shall simply point out that any serious discussion of these matters must also consider the question of whether or not the so-called "orthodox" interpretation of quantum mechanics, which we have been discussing, really provides the best and most comprehensive picture of the physical world. In any case, it seems safe to say that the problem of "determinism in Nature" is no longer considered to be a settled matter, as it was before the invention of quantum mechanics, and indeed, it will probably remain in an unresolved state for some time to come.

4-5 MOTION OF A PARTICLE IN ONE DIMENSION

The foregoing development of the theory of quantum mechanics has been carried out in terms of general observables associated with a general one-dimensional system. We wish now to apply these results to the specific system of a mass m moving along the x-axis in a potential field $V(x)$—a system which we discussed from the standpoint of classical mechanics in Chapter 3. Experiments tell us that our classical treatment of this system is entirely adequate for a "tangible" particle moving over "visible" distances. However, experiments also tell us that our classical description is *not* universally valid; it fails, for example, to correctly describe the behavior of an electron (mass $\simeq 10^{-27}$ gram) on a scale of the order of an atomic diameter (distance $\simeq 10^{-8}$ centimeter). Now, we can expect that the *quantum* treatment of such a system will be valid in both cases; thus we expect that, on the one hand, the quantum description will reduce to the classical description in the *macroscopic* limit, and on the other hand that it will account for such nonclassical phenomena as quantized observables and the wave-particle duality in the *microscopic* limit.

In Sec. 4-5a we shall define and discuss the relevant observable operators for a mass m moving on the x-axis in a potential field $V(x)$. In Sec. 4-5b we shall indicate how these operators lead to a dualistic "wave-particle" behavior. In Sec. 4-5c we shall discuss the way in which the classical description appears as a limiting case of the quantum description. Finally, in Sec. 4-5d we shall work out a simple "quantum mechanics problem" which is typical of those considered in virtually all texts and courses on elementary quantum mechanics.

4-5a Formation of the Observable Operators The Schrödinger Equations and the Position Probability

In classical mechanics, the system consisting of a particle moving along the x-axis has two basic observables—namely, the "position" x and the "momentum" p. Many other observables can be expressed as *functions* of position and momentum; for example, the observables "velocity" and "energy" are given respectively by the functions $v = p/m$ and $E = p^2/2m + V(x)$. Now in quantum mechanics the situation is very much the same: position and momentum are still valid observables, as are also most well-behaved functions of position and momentum. Our last postulate stipulates how we are to form the appropriate *operators* to represent these observables.

Postulate 6. For a particle confined to the x-axis, the observables "position" and "momentum" are represented respectively by the operators

$$\hat{X} = x \tag{4-46}$$

and

$$\hat{P} = -i\hbar \frac{d}{dx} \tag{4-47}$$

Moreover, any observable which in classical mechanics is some well-behaved function of position and momentum, $f(x,p)$, is represented in quantum mechanics by the operator $f(\hat{X},\hat{P})$:

$$\mathcal{A} = f(x,p) \quad \text{implies} \quad \hat{A} = f(\hat{X},\hat{P}) = f\left(x, -i\hbar \frac{d}{dx}\right) \tag{4-48}$$

According to Eqs. (4-46) and (4-47), the position operator $\hat{X}$ and the momentum operator $\hat{P}$ are the operators which transform any given function $\phi(x)$ into the respective functions

$$[\hat{X}\phi(x)] \equiv x \cdot \phi(x)$$

and

$$[\hat{P}\phi(x)] \equiv -ih \cdot \frac{d\phi(x)}{dx}$$

Probably the most important application of the rule (4-48) is the formation of the "energy" or Hamiltonian operator $\hat{H}$. Since the energy in classical mechanics is given by Eq. (3-6b) [see also Eq. (3-8)], then according to Postulate 6 we have

$$\hat{H} = \frac{1}{2m}\hat{P}^2 + V(\hat{X}) \tag{4-49a}$$

or

$$\hat{H} = -\frac{\hbar^2}{2m}\frac{d^2}{dx^2} + V(x) \tag{4-49b}$$

That is, the energy operator $\hat{\text{H}}$ transforms any given function $\phi(x)$ into the function

$$[\hat{\text{H}}\phi(x)] \equiv -\frac{\hbar^2}{2m}\frac{d^2\phi(x)}{dx^2} + V(x)\phi(x)$$

The position, momentum and energy are not the *only* observables which the system under consideration might have in quantum mechanics. There may be several other observables, some of them having *no* analogues in classical mechanics, which characterize certain other attributes of the system. However, we shall confine our discussion here to a description of the system solely in terms of the three observables defined above.

Before examining some of the immediate consequences of the foregoing expressions for $\hat{\text{X}}$, $\hat{\text{P}}$ and $\hat{\text{H}}$, it is well to make certain that these three operators are Hermitian, as required by Postulate 2. We recall that $\hat{\text{A}}$ is said to be an Hermitian operator if and only if $(\phi_1, \hat{\text{A}}\phi_2) = (\hat{\text{A}}\phi_1, \phi_2)$ for any two $\mathcal{H}$-vectors $\phi_1(x)$ and $\phi_2(x)$. It will be left as an exercise for the reader to show that the position operator in Eq. (4-46) satisfies this requirement [see Exercise 49]. To show that the momentum operator in Eq. (4-47) is Hermitian, we proceed as follows: Using the definition of the inner product in Eq. (2-32), we have for any two $\mathcal{H}$-vectors $\phi_1(x)$ and $\phi_2(x)$

$$(\phi_1, \hat{\text{P}}\phi_2) = \int_{-\infty}^{\infty} \phi_1^*(x)\,[\hat{\text{P}}\phi_2(x)]\,dx = \int_{-\infty}^{\infty} \phi_1^*(x)\left[-i\hbar\frac{d\phi_2(x)}{dx}\right]dx$$

$$= -i\hbar \int_{-\infty}^{\infty} \phi_1^*(x)d\,[\phi_2(x)]$$

Integrating by parts, we obtain

$$(\phi_1, \hat{\text{P}}\phi_2) = -i\hbar\left\{\phi_1^*(x)\phi_2(x)\Big|_{-\infty}^{+\infty} - \int_{-\infty}^{\infty} \phi_2(x)d\,[\phi_1^*(x)]\right\}$$

Now, the first term on the right vanishes for the following reason: We proved in Sec. 2-3 that the inner product of any two $\mathcal{H}$-vectors $\phi_1(x)$ and $\phi_2(x)$ exists in the sense that

$$|(\phi_1, \phi_2)| \equiv \left|\int_{-\infty}^{\infty} \phi_1^*(x)\phi_2(x)dx\right| < \infty$$

But this can be true only if both the real and imaginary parts of the integrand, $\phi_1^*(x)\phi_2(x)$, approach zero as $x \to \pm\infty$. Thus, the first term on the right in the previous equation vanishes, and we are left with

$$(\phi_1, \hat{P}\phi_2) = i\hbar \int_{-\infty}^{\infty} \phi_2(x) \left[\frac{d\phi_1^*(x)}{dx}\, dx\right] = i\hbar \int_{-\infty}^{\infty} \left[\frac{d\phi_1(x)}{dx}\right]^* \phi_2(x)\, dx$$

$$= \int_{-\infty}^{\infty} \left[-i\hbar \frac{d\phi_1(x)}{dx}\right]^* \phi_2(x) dx = \int_{-\infty}^{\infty} [\hat{P}\phi_1(x)]^* \phi_2(x)\, dx$$

so

$$(\phi_1, \hat{P}\phi_2) = (\hat{P}\phi_1, \phi_2)$$

which proves that $\hat{P}$ is indeed an Hermitian operator.

Exercise 49.

(a) Prove that the position operator $\hat{X}$, as defined by Eq. (4-46), is an Hermitian operator.

(b) In the same way, show that if $f(x)$ is any well-behaved real function of x, then the operator $f(\hat{X})$ is an Hermitian operator.

Finally, to show that the energy operator H in Eq. (4-49) is Hermitian, we can proceed most simply as follows: We first observe from Eq. (4-49a) that $\hat{H}$ is the *sum* of two operators, namely the kinetic energy operator, $\hat{P}^2/2m$, and the potential energy operator, $V(\hat{X})$. The second operator is Hermitian in consequence of part (b) of the preceding exercise; for the first operator, we note that since $\hat{P}$ is Hermitian, then

$$\left(\phi_1, \frac{1}{2m}\hat{P}^2 \phi_2\right) = \frac{1}{2m}(\phi_1, \hat{P}\hat{P}\phi_2) = \frac{1}{2m}(\hat{P}\phi_1, \hat{P}\phi_2)$$

$$= \frac{1}{2m}(\hat{P}\hat{P}\phi_1, \phi_2) = \left(\frac{1}{2m}\hat{P}^2 \phi_1, \phi_2\right)$$

Therefore, we see that the kinetic energy operator and the potential energy operator are *each* Hermitian; the Hermiticity of $\hat{H}$ then follows from the theorem proved in Exercise 16 that the sum of two Hermitian operators is itself Hermitian.

We have now postulated the precise forms for the position operator $\hat{X}$, the momentum operator $\hat{P}$, and the energy or Hamiltonian operator $\hat{H}$, and we have demonstrated that these operators are Hermitian as required by Postulate 2. Let us next examine some of the important consequences of so representing these observables by these operators. We consider first the energy operator $\hat{H}$.

It will be recalled that the operator $\hat{\text{H}}$ was crucially involved in our statement of Postulate 5. Indeed, with Eq. (4-49b), we find that Eqs. (4-25) and (4-26) take the following respective forms:†

$$-\frac{\hbar^2}{2m}\frac{d^2}{dx^2}\eta_n(x) + V(x)\eta_n(x) = E_n\eta_n(x) \tag{4-50}$$

$$-\frac{\hbar^2}{2m}\frac{\partial^2}{\partial x^2}\Psi(x,t) + V(x)\Psi(x,t) = i\hbar\frac{\partial}{\partial t}\Psi(x,t) \tag{4-51}$$

These two equations are the celebrated Schrödinger equations of quantum mechanics: Eq. (4-50) is called the *time-independent Schrödinger equation*, and Eq. (4-51) is called the *time-dependent Schrödinger equation.* We showed in Sec. 4-4c that if the solutions $\{\eta_n(x)\}$ and $\{E_n\}$ to the first equation can be found, then a general solution $\Psi(x,t)$ to the second equation can be written down immediately, provided $\Psi(x,0)$ is given [see Eq. (4-42)]. However, it is very important, from a logical standpoint, not to confuse these two equations. The time-independent Schrödinger equation (4-50) is the *eigenvalue equation for the energy operator*, whereas the time-dependent Schrödinger equation (4-51) is the *fundamental time-evolution equation for the state vector.* This very basic distinction between the two Schrödinger equations should not become obscured by the similarities in their appearances or by the close relationship between their solutions.

In classical mechanics we know that the motion of a particle in a potential field $V(x)$ is unaltered if we add to $V(x)$ any constant C [see part (a) of Exercise 21]. It is not difficult to show that this is also true in quantum mechanics: If $V(x)$ is replaced by $V'(x) \equiv V(x) + C$, then it follows from Eqs. (4-49) that the new Hamiltonian operator will be

$$\hat{\text{H}}' = \hat{\text{H}} + C$$

where $\hat{\text{H}}$ is the Hamiltonian for $V(x)$. If $\{\eta_n(x)\}$ and $\{E_n\}$ are the eigenvectors and eigenvalues of $\hat{\text{H}}$, then clearly

$$\hat{\text{H}}'\eta_n(x) = \hat{\text{H}}\eta_n(x) + C\eta_n(x) = E_n\eta_n(x) + C\eta_n(x) = (E_n + C)\eta_n(x)$$

from which we may conclude that the eigenvectors and eigenvalues of $\hat{\text{H}}'$ are

$$\left.\begin{aligned} \eta_n'(x) &= \eta_n(x) \\ E_n' &= E_n + C \end{aligned}\right\} n = 1,2,\ldots$$

†In Eq. (4-51) we have written "∂" instead of "d" to emphasize that the x-differentiation of $\Psi(x,t)$ is to be performed treating t as a constant, while the t-differentiation of $\Psi(x,t)$ is to be performed treating x as a constant. See footnote, p. 37.

Now with $\hat{\mathrm{H}}'$ as the Hamiltonian, the system will evolve in time t from any given initial state $\Psi_0(x)$ to the state

$$\Psi'_t(x) = \sum_{n=1}^{\infty} (\eta'_n, \Psi_0) e^{-iE'_n t/\hbar} \eta'_n(x) = \sum_{n=1}^{\infty} (\eta_n, \Psi_0) e^{-i(E_n+C)t/\hbar} \eta_n(x)$$

Using Eq. (2-20c), this is just

$$\Psi'_t(x) = e^{-iCt/\hbar} \sum_{n=1}^{\infty} (\eta_n, \Psi_0) e^{-iE_n t/\hbar} \eta_n(x) = e^{-iCt/\hbar} \Psi_t(x)$$

where $\Psi_t(x)$ is what the state vector would have been if the system's Hamiltonian operator were $\hat{\mathrm{H}}$ instead of $\hat{\mathrm{H}}'$. The essential point here is that $\Psi'_t(x)$ differs from $\Psi_t(x)$ only by a scalar factor of square modulus unity; thus, according to Postulate 1, $\Psi'_t(x)$ and $\Psi_t(x)$ correspond to the same physical state. We see then that, in both classical mechanics and quantum mechanics, an additive constant C in the potential function has no effect on the motion of the system. In classical mechanics this is a consequence of the fact that C, considered as a *function of* x, has zero derivative; in quantum mechanics, on the other hand, this is evidently a consequence of the fact that C, considered as a *Hilbert space operator*, has all functions as eigenfunctions with itself as eigenvalue.

The finding of those functions $\{\eta_n(x)\}$ and numbers $\{E_n\}$ which render Eq. (4-50) an identity—i.e., the *solving* of the time-independent Schrödinger equation—is of special importance. Not only does this yield the physically important "energy levels" of the system, $E_1, E_2, \ldots$, but, as mentioned above, it also provides us with an explicit representation for the time-varying state vector through Eq. (4-42) [or equivalently, Eqs. (4-44) and (4-45)]. Now in order to solve Eq. (4–50), it is clearly necessary to specify a definite form for the potential function, $V(x)$. But having done this, one is then usually faced with a very formidable exercise in the application of the methods and techniques of differential equation theory; indeed, Eq. (4-50) has been solved *exactly* only for a very few simple forms for $V(x)$.

As an example of the kinds of results one can expect, we shall simply exhibit, but *not* derive, the energy eigenvectors and eigenvalues for the case

$$V(x) = \frac{1}{2} kx^2 \qquad (k > 0)$$

The reader will recognize this as the potential function for the "harmonic oscillator"—i.e., for a particle experiencing a spring force

$F(x) \equiv -dV/dx = -kx$, where k is the spring stiffness. Now, in classical mechanics, we know that a solution to Newton's equation (3-3a) for this force function leads to the conclusion that the particle oscillates sinusoidally about the origin with frequency $\nu = \omega/2\pi$, where ω is defined by

$$\omega \equiv \sqrt{k/m}$$

Furthermore, if A is the "amplitude" of these oscillations (i.e., the maximum value of x), then the energy of the system is

$$E_A = \frac{1}{2}kA^2 \equiv \frac{1}{2}m\omega^2 A^2 \qquad \text{(classical harmonic oscillator)} \tag{4-52a}$$

Since A can have any nonnegative value, then according to Eq. (4-52a) the energy of this system in classical mechanics can have *any value* greater than or equal to zero. To examine this problem from the standpoint of quantum mechanics, we must evidently solve the time-independent Schrödinger equation, which in this case takes the form

$$-\frac{\hbar^2}{2m}\frac{d^2}{dx^2}\eta_n(x) + \frac{kx^2}{2}\eta_n(x) = E_n\eta_n(x)$$

In terms of the quantity ω defined above, it is found after considerable mathematical labor that the eigenvalues are

$$E_n = \left(n + \frac{1}{2}\right)\hbar\omega \quad n = 0,1,2,\ldots \text{ (quantum harmonic oscillator)} \tag{4-52b}$$

and the corresponding eigenvectors are given by

$$\eta_n(x) = \frac{(-1)^n}{\sqrt{\sqrt{\pi}\,2^n n!}}\left(\frac{\hbar}{m\omega}\right)^{\frac{2n-1}{4}} e^{\frac{m\omega}{2\hbar}x^2}\frac{d^n}{dx^n}\left[e^{-\frac{m\omega}{\hbar}x^2}\right]$$

We shall not offer any comment here upon the expression for the energy eigenvectors of the harmonic oscillator, except to note that they turn out to be pure real, and also that they can be shown to satisfy the requisite conditions of orthonormality and completeness in Eqs. (4-2). With regard to the energy eigenvalues in Eq. (4-52b), we see that, in contrast to the situation in classical mechanics, the allowed energy values are *discrete* rather than continuous, with a separation between the levels of $\hbar\omega \equiv \hbar\sqrt{k/m}$; moreover, the lowest energy level is *not* zero, but $\hbar\omega/2$. It is also interesting to note that these energy levels are very similar to those which were *postulated* by Planck for the radiating oscillators inside a constant temperature cavity [see Chapter 1].

We shall explore the relation between Eqs. (4-52a) and (4-52b) when we discuss the connection between classical and quantum mechanics in Sec. 4-5c. In Sec. 4-5d we shall examine more closely the details of *solving* the time-independent Schrödinger equation—but for a potential function $V(x)$ which is even simpler than that for the harmonic oscillator. For now, we turn to a consideration of the position and momentum eigenvalue equations.

The treatment of the eigenvalue equations for the position and momentum operators presents problems of a rather peculiar nature. Let x_0 denote an eigenvalue of $\hat{X}$, and let $\delta_{x_0}(x)$ denote the corresponding eigenvector; similarly, let p_0 and $\theta_{p_0}(x)$ denote an eigenvalue and eigenvector of $\hat{P}$:

$$\hat{X}\delta_{x_0}(x) = x_0\,\delta_{x_0}(x)$$

$$\hat{P}\theta_{p_0}(x) = p_0\,\theta_{p_0}(x)$$

Substituting for $\hat{X}$ and $\hat{P}$ their specific forms, these eigenvalue equations read

$$x\,\delta_{x_0}(x) = x_0\,\delta_{x_0}(x) \tag{4-53}$$

$$-i\hbar\frac{d}{dx}\theta_{p_0}(x) = p_0\,\theta_{p_0}(x) \tag{4-54}$$

Exercise 50.

(a) Show from Eq. (4-53) that the function $\delta_{x_0}(x)$ must have the property that $\delta_{x_0}(x) = 0$ for any $x \neq x_0$, but that $\delta_{x_0}(x_0)$ can have *any* value. [*Hint*: Write Eq. (4-53) as $(x - x_0)\delta_{x_0}(x) = 0$.]

(b) Show that the function

$$\theta_{p_0}(x) = e^{ip_0 x/\hbar} \tag{4-55}$$

satisfies Eq. (4-54). [*Hint:* Recall Exercise 7 and Eq. (2-20e).]

Now, it is obvious from Eq. (4-53) that any value for x_0 is as good as any other value; similarly, we see from Eq. (4-55) that p_0 also may have any value. Therefore, the eigenvalues of $\hat{X}$ and $\hat{P}$ are *continuously distributed over the entire real axis:*

$$\left.\begin{array}{l} -\infty < x_0 < +\infty \\ -\infty < p_0 < +\infty \end{array}\right\} \tag{4-56}$$

This is an easily obtained, if not particularly exciting, conclusion about the *eigenvalues* of $\hat{X}$ and $\hat{P}$; however, we run into difficulties when we come to consider the corresponding *eigenvectors.*

With regard to the eigenvector $\delta_{x_0}(x)$, we found in part (a) of Exercise 50 that this function must vanish everywhere except at $x = x_0$, and there it can have any value. Now such a function would normally give zero when computing an expansion coefficient of the kind in Eq. (4-6b).

$$(\delta_{x_0}, \Psi_t) = \int_{-\infty}^{\infty} \delta_{x_0}^*(x)\Psi_t(x)\,dx$$

Indeed, the only way we can *avoid* the unacceptable conclusion that all these expansion coefficients vanish, is to make $\delta_{x_0}(x_0)$ *infinite* in such a way that its product with the infinitesimal dx, $\delta_{x_0}(x_0)\,dx$, is a *finite* number. A rigorous treatment of this highly unusual function lies beyond the scope of this book, and in fact even transcends the scope of ordinary calculus. The function $\delta_{x_0}(x)$ is usually written $\delta(x - x_0)$, and is called by physicists the *Dirac delta function* (although it is not really a "function" in the strict mathematical sense). We shall discuss some of the properties of $\delta(x - x_0)$, as well as some of the important consequences of these properties, in Sec. 4-6b.

At first glance, it might seem that the momentum eigenvectors $\theta_{p_0}(x)$ in Eq. (4-55) are free from any difficulties; however, this is not so. When we compute the norm of $\theta_{p_0}(x)$, we find by virtue of Eq. (2-20d) that

$$(\theta_{p_0}, \theta_{p_0}) = \int_{-\infty}^{\infty} |\theta_{p_0}(x)|^2\,dx = \int_{-\infty}^{\infty} |e^{ip_0x/\hbar}|^2\,dx = \int_{-\infty}^{\infty} 1\cdot dx = \infty$$

That is, there is no way in which the functions $\theta_{p_0}(x)$ can be normalized to unity, as required by Postulates 1 and 2.

The difficulties that we are witnessing with regard to the eigenvectors of the position and momentum operators can ultimately be traced to the fact that these operators have *continuously* distributed eigenvalues [see Eq. (4-56)]. It was precisely to avoid these difficulties that we restricted our discussion in the previous sections to operators with discretely distributed eigenvalues. In order to circumvent these difficulties, it is necessary to modify the definitions of orthonormality and completeness, as these terms apply to eigenbasis vectors associated with continuously distributed eigenvalues. We shall discuss these modifications briefly in Sec. 4-6b; for now, though, all we shall need to know is the following: The *position eigenvector* $\delta_{x_0}(x)$, which corresponds to the eigenvalue x_0, is zero everywhere except at $x = x_0$, at which point it has an "infinite spike"; the *momentum eigenvector* $\theta_{p_0}(x)$, which corresponds to the eigenvalue

p_0, is, apart from some "normalization constant," equal to $\exp(ip_0 x/\hbar)$.

Despite the above difficulties with the eigenvectors of the position and momentum operators, it is possible to calculate the *expectation values* of these observables in a very straightforward manner, by means of the formula in Eq. (4-16):

Exercise 51. Using Eq. (4-16), show that the expectation values of position and momentum in the state $\Psi(x,t)$ are

$$\langle \hat{X} \rangle_t = \int_{-\infty}^{\infty} x\,|\Psi(x,t)|^2\,dx \tag{4-57a}$$

$$\langle \hat{P} \rangle_t = -\,i\hbar \int_{-\infty}^{\infty} \Psi^*(x,t)\,\frac{\partial \Psi(x,t)}{\partial x}\,dx \tag{4-58a}$$

Moreover, the *uncertainties* in position and momentum in the state $\Psi(x,t)$ may be calculated from Eq. (4-14) if we compute, in addition, the quantities $\langle \hat{X}^2 \rangle_t$ and $\langle \hat{P}^2 \rangle_t$:

Exercise 52. Show that

$$\langle \hat{X}^2 \rangle_t = \int_{-\infty}^{\infty} x^2\,|\Psi(x,t)|^2\,dx \tag{4-57b}$$

$$\langle \hat{P}^2 \rangle_t = \hbar^2 \int_{-\infty}^{\infty} \left| \frac{\partial \Psi(x,t)}{\partial x} \right|^2 dx \tag{4-58b}$$

[*Hint:* For $\langle \hat{P}^2 \rangle_t$, use the fact that, since $\hat{P}$ is Hermitian, then $(\Psi_t, \hat{P}^2 \Psi_t) = (\hat{P}\Psi_t, \hat{P}\Psi_t)$.]

In particular, Eqs. (4-57) can easily be generalized: If $f(x)$ is any well-behaved, real function of x, then according to Postulate 6, $f(x)$ is an *observable* with *operator* $f(\hat{X}) = f(x)$. The expectation value of this observable in the state $\Psi(x,t)$ is, according to Eq. (4-17),

$$\langle f(\hat{X}) \rangle_t = (\Psi_t, f(\hat{X})\Psi_t) = \int_{-\infty}^{\infty} \Psi^*(x,t)\,[f(x)\Psi(x,t)]\,dx$$

or

$$\langle f(\hat{X}) \rangle_t = \int_{-\infty}^{\infty} f(x)|\Psi(x,t)|^2\,dx \tag{4-59a}$$

This last equation can be given a very interesting and useful interpretation. We recall from Exercise 26 that, if f has a Taylor series expansion, then the operator $f(\hat{A})$ has eigenvalues $\{f(A_n)\}$, where $\{A_n\}$ are the eigenvalues of $\hat{A}$. Therefore, since $\hat{X}$ has eigenvalues $\{x\}$ for all $-\infty < x < \infty$, then the operator $f(\hat{X})$ has eigenvalues $\{f(x)\}$ for all $-\infty < x < \infty$. Thus, Eq. (4-59a) expresses $\langle f(\hat{X})\rangle_t$ as a sort of "weighted sum" of the possible $f(\hat{X})$-values. To bring this out more explicitly, let us for the moment write Eq. (4-59a) as a *discrete* sum: We partition the entire x-axis into subintervals $\Delta x_1, \Delta x_2, \ldots,$ and we let x_k denote an x-value inside the subinterval Δx_k. Then if we make the lengths of these subintervals infinitesimally small, we may write by the very definition of the integral,

$$\langle f(\hat{X})\rangle_t = \sum_k f(x_k)|\Psi(x_k,t)|^2 \,\Delta x_k \tag{4-59b}$$

We now compare this with Eq. (2-8):

$$\langle f(v)\rangle = \sum_k f(v_k)p_k \qquad [2\text{-}8]$$

which gives the mean value of $f(v)$ for a series of v_k-values distributed with probabilities p_k. Equation (2-8) gives $\langle f(v)\rangle$ as a weighted sum of the possible values $f(v_k)$, while Eq. (4-59b) gives $\langle f(\hat{X})\rangle_t$ as a weighted sum of the possible values $f(x_k)$. Because f is an arbitrary function, these considerations imply the correspondence

$$p_k \longleftrightarrow |\Psi(x_k,t)|^2 \,\Delta x_k \tag{4-60}$$

We note in particular that this correspondence is consistent with the condition $\Sigma_k p_k = 1$, since

$$\sum_k |\Psi(x_k,t)|^2 \,\Delta x_k = \int_{-\infty}^{\infty} |\Psi(x,t)|^2 \,dx = (\Psi_t, \Psi_t) = 1$$

Now Eq. (2-8) gives the mean value of $f(v)$ *because* p_k is the probability that the particular value v_k will be obtained in a random selection from the set of v-values; thus, the correspondence (4-60) *implies* that $|\Psi(x_k,t)|^2\Delta x_k$ is the probability that a value in the particular interval Δx_k will be obtained in a position measurement on the state $\Psi(x,t)$. Returning to the integral expression for $\langle f(\hat{X})\rangle_t$ in Eq. (4-59a), we conclude that:

$|\Psi(x,t)|^2 \,dx$ = the probability that a position measurement on the state $\Psi(x,t)$ will yield a value between x and $x + dx$.

(4-61)

It follows from this that the probability for a position measurement on the state $\Psi(x,t)$ to yield a value somewhere between x_1 and x_2 is

$$P(x_1,x_2;t) = \int_{x_1}^{x_2} |\Psi(x,t)|^2 \, dx \tag{4-62}$$

because this is just the "sum" of the probabilities for measuring a value in any of the dx-intervals between x_1 and x_2.

On account of Eq. (4-61), the quantity $|\Psi(x,t)|^2 \equiv \Psi_t^*(x)\Psi_t(x)$ is called the *position probability density function* (the term "density" is used because Eq. (4-61) implies that $|\Psi(x,t)|^2$ has units of probability *per unit* x). Although the state vector itself has no direct physical significance, we see that its square modulus has a very deep physical significance. This particular aspect of the state vector was first recognized by Max Born, and is undoubtedly one of its most important and frequently used properties. However, it must be emphasized that Eq. (4-61) by no means exhausts all the physical implications of the state vector.

In order to understand fully the significance of the position probability density function, we show in Fig. 4 a plot of $|\Psi(x,t)|^2$ versus x for some hypothetical state $\Psi(x,t)$. If one were to make a series of very many *repeated* measurements of the position on the state $\Psi(x,t)$, and if the results were plotted as a frequency bar graph over small, equal-size bins, then Eq. (4-61) implies that the shape of this graph would follow the shape of the curve in Fig. 4, to within random statistical fluctuations. Thus the curve in Fig. 4 is essentially the same as the curve in Fig. 3, taking into account the fact that the eigenvalues of $\hat{X}$ are continuously distributed over the entire real axis. Since

$$\int_{-\infty}^{\infty} |\Psi(x,t)|^2 \, dx = (\Psi_t,\Psi_t) = 1$$

then the area under the curve in Fig. 4 is unity.

Equation (4-62) implies that the area under the curve $|\Psi(x,t)|^2$ between x_1 and x_2, shown shaded in Fig. 4, is equal to $P(x_1,x_2;t)$, the probability of finding the particle between x_1 and x_2. We have used the phrase "finding the particle" in a special sense, and it is important that we understand precisely what we mean, and what we do not mean, by it. According to our discussion in Sec. 4-3b, *prior* to the position measurement we should *not* try to picture the particle as "really" being either inside or outside the interval $[x_1,x_2]$, with the position measurement "discovering" which of these two alternatives

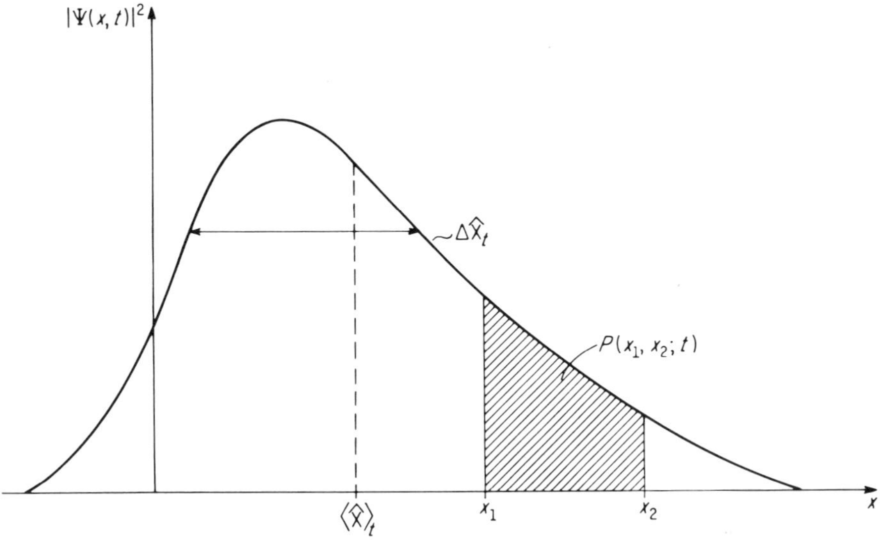

Fig. 4. A plot of the position probability density function $|\Psi(x,t)|^2$ versus x for a hypothetical state $\Psi(x,t)$. The position expectation value and uncertainty are indicated schematically. The total area under the curve is unity, by virtue of the fact that $(\Psi_t, \Psi_t) = 1$; the shaded area under the curve between x_1 and x_2 is numerically equal to $P(x_1, x_2; t)$, the probability that a position measurement at time t will yield a value between x_1 and x_2. We have drawn the curve as though $|\Psi(x,t)|^2$ were fairly well localized in one region of the x-axis; however, it is important to realize that the curve could very well consist of two or more widely separated humps. In such a case, $\Delta\hat{X}_t$ would be large, and $\langle\hat{X}\rangle_t$ would not be a particularly useful quantity.

is actually the case. For, if the state vector $\Psi_t(x)$ does *not* coincide with one of the position eigenvectors $\delta_{x_0}(x)$—as is obviously the case in Fig. 4—then we may *not* speak of the particle as "having a position" in the usual sense of this phrase. So when we say that $P(x_1, x_2; t)$ is "the probability of finding the particle somewhere between x_1 and x_2," what we really mean is that $P(x_1, x_2; t)$ is "the probability that a position measurement on the state $\Psi(x,t)$ will *develop* a position value for the particle somewhere between x_1 and x_2." It is all right to use the former phrase for brevity, provided we keep in mind the latter, more explicit interpretation.

As time evolves, the state vector $\Psi_t(x)$ changes, and the position probability density curve in Fig. 4 will change its shape and position in some more or less complicated way, subject to the condition that the area under it remain equal to unity. In Sec. 4-5c we shall examine the behavior in time of $\langle\hat{X}\rangle_t$; for now, however, let us examine the time variation of $P(x_1,x_2;t)$. To this end, we first derive a formula for the time-rate-of-change of $P(x_1,x_2;t)$. We have from Eq. (4-62)

$$\frac{d}{dt}P(x_1,x_2;t) = \frac{d}{dt}\int_{x_1}^{x_2} \Psi^*(x,t)\Psi(x,t)\,dx$$

$$= \int_{x_1}^{x_2} \left[\Psi^* \frac{\partial \Psi}{\partial t} + \Psi \frac{\partial \Psi^*}{\partial t}\right] dx$$

Using the time-dependent Schrödinger equation together with its complex conjugate [note that $V(x)$ is pure real],

$$i\hbar \frac{\partial \Psi}{\partial t} = -\frac{\hbar^2}{2m}\frac{\partial^2 \Psi}{\partial x^2} + V\Psi$$

$$-i\hbar \frac{\partial \Psi^*}{\partial t} = -\frac{\hbar^2}{2m}\frac{\partial^2 \Psi^*}{\partial x^2} + V\Psi^*$$

it is not too difficult to eliminate the time derivatives in the above equation to obtain

$$\frac{d}{dt}P(x_1,x_2;t) = \frac{i\hbar}{2m}\int_{x_1}^{x_2}\left[\Psi^*\frac{\partial^2 \Psi}{\partial x^2} - \Psi\frac{\partial^2 \Psi^*}{\partial x^2}\right] dx \qquad (4\text{-}63)$$

Exercise 53. Carry out the steps leading to Eq. (4-63).

The integral on the right-hand side of Eq. (4-63) can be explicitly evaluated by an integration-by-parts. Remembering that "$\partial\Psi/\partial x$" just means "$d\Psi/dx$ with t treated as a constant," we have for the first term in Eq. (4-63)

$$\int_{x_1}^{x_2} \Psi^*\frac{\partial^2 \Psi}{\partial x^2}dx = \int_{x_1}^{x_2}\Psi^*\frac{\partial}{\partial x}\left(\frac{\partial \Psi}{\partial x}\right)dx = \int_{x_1}^{x_2}\Psi^* d\left(\frac{\partial \Psi}{\partial x}\right)$$

$$= \Psi^*\frac{\partial \Psi}{\partial x}\Big|_{x_1}^{x_2} - \int_{x_1}^{x_2}\frac{\partial \Psi}{\partial x}d(\Psi^*)$$

so

$$\int_{x_1}^{x_2}\Psi^*\frac{\partial^2 \Psi}{\partial x^2}dx = \Psi^*\frac{\partial \Psi}{\partial x}\Big|_{x_1}^{x_2} - \int_{x_1}^{x_2}\frac{\partial \Psi}{\partial x}\frac{\partial \Psi^*}{\partial x}dx$$

A similar expression holds for the second term in Eq. (4-63), except that the roles of Ψ and Ψ^* are interchanged. When the expression for the second term is subtracted from that for the first term, the integrals cancel, and we are left with

$$\frac{d}{dt} P(x_1,x_2;t) = \frac{i\hbar}{2m} \left[\Psi^* \frac{\partial \Psi}{\partial x} \bigg|_{x_1}^{x_2} - \Psi \frac{\partial \Psi^*}{\partial x} \bigg|_{x_1}^{x_2} \right]$$

Defining now the quantity $S(x,t)$ by

$$S(x,t) \equiv -\frac{i\hbar}{2m} \left(\Psi^*(x,t) \frac{\partial \Psi(x,t)}{\partial x} - \Psi(x,t) \frac{\partial \Psi^*(x,t)}{\partial x} \right) \qquad (4\text{-}64)$$

we obtain the final result

$$\frac{d}{dt} P(x_1,x_2;t) = S(x_1,t) - S(x_2,t) \qquad (4\text{-}65)$$

In order to give a physical interpretation to this result, and in particular to the quantity $S(x,t)$, let us for the moment regard Eq. (4-65) simply as a formula for the time-rate-of-change of the shaded area in Fig. 4. Since the *total* area under the curve in Fig. 4 is always equal to unity, then any increase in the area inside $[x_1,x_2]$ must be accompanied by an equal decrease in the area outside $[x_1,x_2]$, and vice versa. Thus, it is reasonable to think of the change in the shaded area in Fig. 4 as resulting from a "flow of area" across the two boundary lines at x_1 and x_2. More specifically, suppose we let $R(x,t)$ denote the *rate* at which area is crossing the point x at time t in the positive x-direction, with the convention that an area flow in the negative x-direction is specified by a negative value for $R(x,t)$. The quantity $R(x_1,t) + (-R(x_2,t))$ would then denote the rate at which area is entering the interval $[x_1,x_2]$ at x_1 plus the rate at which area is entering this interval at x_2; clearly, this is just the *net* rate of increase of the area inside $[x_1,x_2]$, $dP(x_1,x_2;t)/dt$. We now observe that Eq. (4-65) is precisely of this form; consequently, we may interpret the quantity $S(x,t)$ as being the instantaneous rate at which area or "position probability" is crossing the point x in the positive x-direction, with negative values for $S(x,t)$ signifying a flow of position probability in the negative x-direction. We call $S(x,t)$ the *position probability current* at the point x at time t.

With this interpretation of $S(x,t)$, Eq. (4-65) is merely the statement that the instantaneous rate of change of the probability for finding the particle between x_1 and x_2, is determined solely by the instantaneous values of the position probability currents at x_1 and x_2; if the current at x_1 is larger (smaller) than the current at x_2, then the probability of finding the particle between x_1 and x_2 is increasing (decreasing).

Exercise 54.

(a) With the help of Eq. (2-13), show that the position probability current can also be written

$$S(x,t) = \frac{\hbar}{m} \operatorname{Im} \left(\Psi^*(x,t) \frac{\partial \Psi(x,t)}{\partial x} \right) \tag{4-66}$$

which, incidently, shows that $S(x,t)$ is *pure real*, as we would expect.

(b) Suppose the potential function $V(x)$ is such that the energy eigenvectors $\{\eta_n(x)\}$ turn out to be pure real. For this case, show that the position probability current vanishes identically if the system is in a stationary state, $\Psi^{(n)}(x,t)$.

In most practical situations, the physicist deals with a large number N of noninteracting particles, all subject to the same potential and all in the same state $\Psi(x,t)$. In such a case, he often speaks of there being $N \cdot P(x_1,x_2;t)$ particles inside the interval $[x_1,x_2]$, and $N \cdot S(x,t)$ particles per second crossing the point x in the positive x-direction (or negative x-direction if $S(x,t) < 0$). Although these statements are *not* literally correct within the framework of orthodox quantum mechanics—for example, they would be meaningless for $N = 1$—no *practical* difficulty is encountered if one does not try to single out *specific ones* of the particles as really being inside $[x_1,x_2]$, or *specific ones* of the particles as really crossing the point x at time t. But strictly speaking, $|\Psi(x,t)|^2$ is a *position probability* density, not a *particle* density, and $S(x,t)$ is a position probability current, not a particle current. These considerations are illustrative of the dramatic revision which quantum mechanics has effected with respect to the familiar, classical concept of a "physical particle." In the next section we shall pursue this conceptual revision to what might be called its logical extreme; in the section following that, we shall show how quantum mechanics allows us to regain, in the *macroscopic limit*, our familiar classical description of a "particle" and its dynamical behavior.

4-5b The Position-Momentum Uncertainty Relation The Wave-Particle Duality

We shall now demonstrate that the postulates of quantum mechanics imply that a particle moving along the x-axis can exhibit the attributes of either a *particle* or a *wave*. In order to avoid confusing the "particle" of our system with the particle "property" or "attribute," we shall adopt for our system an *electron* on the x-axis.

Suppose first that the state vector of the electron coincides with one of the eigenvectors $\delta_{x_0}(x)$ of the position operator $\hat{X}$. According to our discussion in Sec. 4-3b, we may say in this case—and *only* in this case—that the observable *position* "has the value x_0," or more simply that the electron "is at the point x_0." Note that this conclusion is entirely consistent with the *spatially localized* form of the function $\delta_{x_0}(x)$: since this function has an infinite spike at $x = x_0$ and vanishes everywhere else, then the position probability density function $|\delta_{x_0}(x)|^2$ clearly implies that a measurement of position will necessarily find the electron at the point $x = x_0$. Now it must be emphasized that this property of "having a position" or of "being spatially localized" is essentially the *defining* property of a *particle*. And it is *only* when the state vector of the electron coincides with one of these infinitely localized position eigenvectors $\delta_{x_0}(x)$ that we can meaningfully assert that the electron "has a position" and therefore "is a particle."

Suppose, on the other hand, that the state vector of the electron coincides with one of the eigenvectors $\theta_{p_0}(x)$ of the momentum operator $\hat{P}$. According to our discussion in Sec. 4-3b, we may say in this case—and *only* in this case—that the observable *momentum* "has the value p_0," or more simply that the electron "is moving with momentum p_0." Now, if we write down the explicit form of the function $\theta_{p_0}(x)$ in Eq. (4-55) [see Eq. (2-20a)],

$$\theta_{p_0}(x) = e^{ip_0x/\hbar} \equiv \cos\left[\frac{p_0 x}{\hbar}\right] + i \sin\left[\frac{p_0 x}{\hbar}\right]$$

we can see that this function is definitely *not* localized on the x-axis. Instead, $\theta_{p_0}(x)$ is seen to have an *infinite, periodic, spatial extension*, with fundamental period or "wavelength"

$$\lambda_0 = h/p_0 \tag{4-67}$$

Exercise 55. Show that $\theta_{p_0}(x)$ is periodic in x with period $\lambda_0 = h/p_0$. [*Hint*: Prove that $\theta_{p_0}(x + \lambda_0) \equiv \theta_{p_0}(x)$.]

Now the property of *periodic spatial extension* is essentially the defining attribute of a *wave*, just as the property of *sharp spatial localization* is the defining attribute of a *particle*. Therefore, when the state vector of the electron coincides with the momentum eigenvector $\theta_{p_0}(x)$—i.e., when the electron has momentum p_0—then the electron is in a certain sense a *wave*, and can be said to "have a wavelength" whose numerical value is given by Eq. (4-67). We note in passing that this is essentially the same conclusion that physicists were *experimentally* led to *before* the invention of quantum mechanics [recall our discussion of Eq. (1-2)].†

We may summarize these results by saying that, when the state vector of the electron coincides with an eigenvector of $\hat{X}$, then the electron has the attributes of (and therefore "is") a *particle*; on the other hand, when the state vector of the electron coincides with an eigenvector of $\hat{P}$, then the electron has the attributes of (and therefore "is") a *wave*. The questions now arise, if the electron is in the state $\delta_{x_0}(x)$, does it have any wave attributes, and if the electron is in the state $\theta_{p_0}(x)$, does it have any particle attributes? Logically, we expect a negative answer to both questions, since the properties of "sharp spatial localization" and "periodic spatial extension" are mutually exclusive properties.

To verify this conjecture, suppose first that the electron is in the state $\Psi_t(x) = \delta_{x_0}(x)$, and suppose we wish to calculate the expectation value of the momentum. Following Eq. (4-58a), we write

$$\langle\hat{P}\rangle_t = -i\hbar \int_{-\infty}^{\infty} \delta_{x_0}^*(x) \frac{d\delta_{x_0}(x)}{dx} dx$$

$$= -i\hbar \int_{-\infty}^{\infty} \left(\frac{d\delta_{x_0}(x)}{dx}\right) (\delta_{x_0}(x) \cdot dx)$$

where we have used the fact that $\delta_{x_0}(x)$ is real. Now, since $\delta_{x_0}(x) = 0$ for all $x \neq x_0$, the only contribution to the integral comes at $x = x_0$; here, the product $\delta_{x_0} \cdot dx$ is *finite*, but $d\delta_{x_0}/dx$ is *undefined* owing to the radical discontinuity in $\delta_{x_0}(x)$ at $x = x_0$. Thus, $\langle P \rangle_t$ is essentially *undefined* in the state $\Psi_t(x) = \delta_{x_0}(x)$. Let us suppose next that the electron is in the state $\Psi_t(x) = \theta_{p_0}(x)$, and suppose we wish to calculate the position probability density function. According to Eqs. (4-61) and (2-20d), we have

$$|\Psi_t(x)|^2 = |\theta_{p_0}(x)|^2 = |e^{ip_0 x/\hbar}|^2 = 1$$

†It should be pointed out that, although $\theta_{p_0}(x)$ is periodic in x, the corresponding position probability density function, $|\theta_{p_0}(x)|^2$, is *not* periodic: as we shall see shortly, the square modulus of $\theta_{p_0}(x)$ is simply a constant. However, the periodic nature of $\theta_{p_0}(x)$ can be rendered "physically observable" by causing a beam of electrons with momentum p_0 to *interact* with a suitable apparatus, such as the crystal diffraction grating of the Davisson-Germer experiment. An analysis of this "interaction," which in itself is essentially a "measurement" of momentum or wavelength, is too complicated for us to consider here.

But this implies that a position measurement is *equally* likely to yield *any* value for x.

These unusual conclusions may best be understood with the help of the Compatibility Theorem and the Heisenberg Uncertainty Principle. To this end, we first make the following simple calculation:

Exercise 56. Prove that the operators $\hat{X}$ and $\hat{P}$, as defined in Postulate 6, satisfy the relation

$$\hat{X}\hat{P} - \hat{P}\hat{X} = i\hbar \tag{4-68}$$

[*Hint*: Prove that, for any $\mathcal{H}$-vector $\phi(x)$, $\hat{X}[\hat{P}\phi(x)] - \hat{P}[\hat{X}\phi(x)] = i\hbar\phi(x)$.]

So the position and momentum operators *do not commute*. Therefore, the Compatibility Theorem tells us that position and momentum are *not* "compatible" or "simultaneously measureable." Moreover, upon substituting Eq. (4-68) into the Heisenberg Uncertainty Relation, we find

$$\Delta\hat{X}_t \cdot \Delta\hat{P}_t \geqq \frac{\hbar}{2} \tag{4-69}$$

Exercise 57. Derive Eq. (4-69).

The above inequality is known as the position-momentum uncertainty relation; it is obviously very similar in form to the time-energy uncertainty relation in Eq. (4-36), although it must be noted that the quantity T_a in Eq. (4-36) is *not* to be regarded as an uncertainty in some sort of "time operator." According to Eq. (4-69), the more precisely the position of the electron is defined (i.e., the smaller $\Delta\hat{X}_t$ is), the less precisely the momentum of the electron is defined (i.e., the larger ΔP_t must be)—and vice versa. Indeed, if the electron can be said to "have a position" (i.e., if $\Delta\hat{X}_t = 0$), then it *cannot* be said to "have a momentum" (i.e., $\Delta\hat{P}_t$ must be infinite)—and of course vice versa. Therefore we see that, although quantum mechanics allows an electron to possess "particle" attributes *and* "wave" attributes, it expressly *forbids* the electron from being a particle and a wave *simultaneously*.

In light of these conclusions, we can understand the so-called wave-particle duality of Nature by the following chain of reasoning:

If we measure the *position* of an electron, then regardless of the state vector of the electron just prior to the measurement, immediately *after* the measurement it will coincide with one of the position eigenvectors, $\delta_{x_0}(x)$ [by Postulate 4]. We will then have $\Delta\hat{X} = 0$ [by Exercise 33], so that the position of the electron is well-defined and can be said to "have a value." Evidently, the position measurement has *endowed* the state vector of the electron with the property of *sharp spatial localization*, so that the electron may truly be regarded as a "particle." However, the electron then has *no* wavelike attributes: for since $\Delta\hat{X} = 0$ then Eq. (4-69) requires that $\Delta\hat{P} = \infty$, which means that the electron cannot be said to "have a momentum or wavelength." Thus, the *measurement* of the observable "position" has *developed* the particle nature of the electron, but it has at the same time *destroyed* the wave nature of the electron.

Exercise 58. Rewrite the preceding paragraph, except start out with the phrase, "If we measure the *momentum* or *wavelength* of an electron, then . . ."

The wave-particle duality of Nature, when viewed from the standpoint of classical mechanics, represented a genuine paradox; however, it should be clear from our discussion here that this phenomenon emerges as a very logical consequence of the basic tenets of quantum mechanics. Historically, the duality was one of the chief motivations for seeking an alternative to classical mechanics. From a modern point of view, the existence in Nature of the wave-particle duality provides strong evidence for the validity of the entire quantum theory.

The real source of the wave-particle duality is evidently the strict incompatibility of position and momentum. A full appreciation of this incompatibility is essential for a proper understanding of many results in quantum mechanics. The following exercise provides a case in point.

Exercise 59. Consider a particle of mass m in an harmonic oscillator potential, $V(x) = kx^2/2 \equiv m\omega^2 x^2/2$.

(a) Suppose the particle is a *classical* particle with total energy E. Show that a position measurement *cannot* find the particle outside the interval $-\sqrt{2E/m\omega^2} \leqq x \leqq \sqrt{2E/m\omega^2}$. [*Hint*: Recall the discussion of Fig. 2.]

(b) Suppose the particle is a *quantum* particle with total energy E_n. Show that a position measurement *can* find the particle outside the interval $-\sqrt{2E_n/m\omega^2} \leqq x \leqq \sqrt{2E_n/m\omega^2}$. [*Hint*: Since the particle has total energy E_n, it must be in the stationary state $\eta_n(x) \exp(-iE_n t/\hbar)$, where $\eta_n(x)$ and E_n were written down for the harmonic oscillator in Sec. 4-5a. Show, without performing any detailed calculations, that the position probability density function for this stationary state does not vanish identically outside the interval in question.]

The phenomenon described in part (b) of the above exercise is not peculiar to the harmonic oscillator potential alone; it is found to occur for nearly all potentials of a similar "concave-up" shape. From a classical point of view, however, this phenomenon seems quite paradoxical: if the particle can be found at points where $V(x) > E$, does not this imply the patently untenable conclusion that the kinetic energy is negative? To understand why this reasoning is fallacious from a quantum viewpoint, let us consider the five observables, position, momentum, kinetic energy, potential energy, and total energy. Of the five corresponding operators, $\hat{X}$, $\hat{P}$, $\hat{P}^2/2m$, $V(\hat{X})$, and $\hat{P}^2/2m + V(\hat{X})$, straightforward calculations utilizing Eq. (4-68) reveal that two and *only* two pairs commute; more specifically the only commuting pairs of operators are $\hat{P}$ and $\hat{P}^2/2m$, and, for any reasonable function V, $\hat{X}$ and $V(\hat{X})$ [see Exercise 36]. It follows from the Compatibility Theorem that momentum and kinetic energy are compatible, as are also position and potential energy; however, position and kinetic energy are *not* compatible, momentum and potential energy are *not* compatible, and the total energy is *not* compatible with *any* of the other observables.† To verify these incompatibilities, we need only observe the great dissimilarities among the position/potential energy eigenvectors $\{\delta_{x_0}(x)\}$, the momentum/kinetic energy eigenvectors $\{\theta_{p_0}(x)\}$, and, for the harmonic oscillator, the total energy eigenvectors $\{\eta_n(x)\}$. In view of these incompatibilities we may evidently refute the classical objection, that an harmonic oscillator would have to have a negative kinetic energy if its position were such that its potential energy exceeded its total energy, simply by observing that the oscillator *cannot* be said to "have values" for all these variables simultaneously. If the oscillator "has energy E_n," it *cannot* sensibly be said to "have values" of position and kinetic energy, because $\eta_n(x)$ is not an eigenvector of either the position operator or the kinetic energy operator; moreover, if the position is *measured*, the state vector will thereby be forced into one of the position eigenvectors $\delta_{x_0}(x)$, and it will *then* be impossible to ascribe to the oscillator either a kinetic energy or a total energy. Thus we see that, because of the fundamental incompatibility between position and momentum, an harmonic oscillator can have *either* a position *or* a kinetic energy *or* a total energy, but *not* any two of these simultaneously; moreover, whatever value may actually be realized for one of these observables at any particular time will necessarily be one of the legitimate eigenvalues of the respective operator—namely, x_0 or $p_0^2/2m$ or $(n + 1/2)\hbar\omega$.

†An exception to this last statement arises if $V(x)$ = constant, in which case the kinetic and total energies are essentially the same.

As the reader pursues his study of quantum mechanics he will constantly feel the urge to try to understand or "see through" quantum phenomena in terms of common-sense classical concepts. Most of the nonorthodox views of quantum mechanics try, in varying degrees, to do just this. The orthodox view, which we have taken here, must be regarded as rather radical in this respect: it asserts, or at least strongly suggests, that there *is no* adequate interpretation of quantum mechanics via purely classical notions, and that the message which Nature is trying to get across to us is simply that such classical ways of thinking do not apply to the microscopic physical world. If we accept this view, then we would evidently be better employed in trying to "see through" classical mechanics in terms of the concepts of quantum mechanics. This, in short, will be our goal in the next section.

4-5c The Ehrenfest Equations The Classical Limit of Quantum Mechanics

We have seen that quantum mechanics must be used instead of classical mechanics when one deals with the *microscopic* physical world. However, this does not mean that we should discard the classical theory altogether; for if we can be certain of anything at all in physics, it is that classical mechanics correctly and efficiently describes many aspects of the *macroscopic* physical world. Therefore, if quantum mechanics is indeed a more comprehensive theory of physical phenomena than classical mechanics, then it is incumbent upon the former to *reduce to* rather than replace the latter in the macroscopic limit.

In order to demonstrate that quantum mechanics does satisfy this requirement, it is sufficient to show that, for any observable $\mathcal{A}$ which has a classical analogue:

(i) In the macroscopic limit, any discreteness in the eigenvalues of $\hat{A}$ is not noticeable.

(ii) In the macroscopic limit, the uncertainty in $\mathcal{A}$, $\Delta\hat{A}_t$, is in practice so small in comparison with the expectation value of $\mathcal{A}$, $\langle\hat{A}\rangle_t$, that we can for all practical purposes say that "$\mathcal{A}$ has the value $\langle\hat{A}\rangle_t$."

(iii) In the macroscopic limit, the time evolution equation for $\langle\hat{A}\rangle_t$ coincides with the classical equation of motion for $\mathcal{A}(t)$.

Clearly, if these three requirements are fulfilled, then a measurement of $\mathcal{A}$ will in effect *always* yield the value $\langle\hat{A}\rangle_t$, and moreover this "value of $\mathcal{Q}$" will evolve continuously with time according to the laws of classical mechanics.

Unfortunately, we shall not be able to give rigorous, general proofs of the first two requirements; however, we will at least be able to see that they are quite plausible. Let us consider first requirement (i). We have seen in Sec. 4-5a that the position and momentum operators already possess continuously distributed eigenvalues; however, the energy operator is often found to have discrete eigenvalues. For example, in Sec. 4-5a we mentioned (without proof) that for the linear harmonic oscillator potential, $V(x) = kx^2/2 \equiv m\omega^2 x^2/2$, the energy eigenvalues are given by [see Eq. (4-52b)]

$$E_n = \left(n + \frac{1}{2}\right)\hbar\omega \qquad n = 0,1,2,\ldots$$

Therefore, the *relative spacing* of the energy levels around the value E_n is

$$\frac{E_{n+1} - E_n}{E_n} = \frac{\hbar\omega}{\left(n + \frac{1}{2}\right)\hbar\omega} = \frac{1}{n + \frac{1}{2}}$$

Evidently, this relative spacing will be small if n is large, in which case it is approximately $1/n$. Now, *classically* such a particle would oscillate sinusoidally about the origin with frequency

$$\nu = \omega/2\pi$$

and could do so with any of the energies [see Eq. (4-52a)]

$$E_A = \frac{1}{2}m\omega^2 A^2 \qquad A \geqq 0$$

Now in the *macroscopic limit* we know that this last equation is essentially valid; since the quantum expression is presumed to be universally valid, then the quantum number n must be such that

$$\left.\begin{aligned} &E_n \cong E_A \\ &\text{or} \\ &\left(n + \frac{1}{2}\right) \cong \frac{1}{2}\left(\frac{m}{\hbar}\right)\omega A^2 \end{aligned}\right\} \text{macroscopic limit}$$

Exercise 60.

(a) Show that, for a 1 gram particle oscillating with frequency 3 cycles per second and amplitude 1 centimeter, n is roughly 10^{28}. Thus, conclude that the relative spacing between neighboring quantum energy levels is so small that it could never be detected experimentally. [Use $\hbar = 1.054 \times 10^{-27}$ erg · sec.]

(b) Repeat this calculation for an electron ($m \simeq 10^{-27}$ gram) oscillating with a frequency on the order of visible light ($\nu \simeq 10^{15}$ cps) and amplitude on the order of an atomic diameter ($A \simeq 10^{-8}$ cm). Would quantum effects be noticeable in this case?

(c) For the system of part (a), show that the lowest energy level, corresponding to $n = 0$, is so small on the macroscopic scale that it could not be experimentally distinguished from the classical minimum of zero.

The results of the preceding exercise are in many respects typical of most potentials encountered in quantum mechanics: Owing to the smallness of $\hbar$, macroscopic energies necessarily correspond to large values of the quantum number n, and this in turn implies a relative spacing between the nearby levels which is so small that the discrete values *appear* to be continuous. Only when one enters the atomic or subatomic realms does the spacing between the levels become significant.

We consider next requirement (ii). We found in the preceding section that position and momentum are *not* compatible observables; in fact, according to Eq. (4-69),

$$\Delta \hat{X}_t \cdot \Delta \hat{P}_t \geqq \frac{\hbar}{2}$$

so it impossible for position and momentum to simultaneously have *exactly* defined values. However, let us get some idea of just how stringent this limitation is from a *macroscopic* point of view. Note first that if the particle has a momentum uncertainty of $\Delta \hat{P}_t$, then it will have a velocity uncertainty of $\Delta \hat{V}_t = \Delta \hat{P}_t / m$.† Therefore the uncertainties in position and velocity must always satisfy

$$\Delta \hat{X}_t \cdot \Delta \hat{V}_t \geqq \frac{\hbar}{2m} \tag{4-70}$$

Exercise 61.

(a) Show that a 1-gram particle can have its position defined to within 0.001 micron and its velocity defined to within 0.001 micron per century, and yet the uncertainty relation in Eq. (4-70) would *not* be violated. [*Note*: 1 micron = 10^{-4} cm, and 1 year $\cong 3 \times 10^7$ sec.]

†Since $v = p/m$ in classical mechanics, then Postulate 6 implies that the quantum velocity operator is just $\hat{V} = \frac{1}{m}\hat{P}$.

(b) For an electron (mass $\simeq 10^{-27}$ gram) confined to an interval on the order of an atomic diameter (distance $\simeq 10^{-8}$ cm), what is the *minimum* value for the uncertainty in the velocity? Would it make sense in this case to speak of the velocity of the electron as "having a value?" [Note for comparison that 1 cm/sec is roughly equal to 2×10^{-2} mile per hour.]

We can see from part (a) of this exercise that it is quite possible, at least from the standpoint of the Uncertainty Principle alone, for a *macroscopic* particle to have its position and its momentum *simultaneously* defined with great precision. Again, we see that the reason for this is the extreme smallness of the constant $\hbar$: Since $\hbar$ is practically zero on the macroscopic scale, then Eq. (4-68) implies that $\hat{X}$ and $\hat{P}$ approximately commute, so that position and momentum are approximately compatible. It is thus possible for the state of a macroscopic particle to be such that the "widths" of the position and momentum distribution curves, $\Delta\hat{X}_t$ and $\Delta\hat{P}_t$, are *simultaneously so small* in comparison with macroscopic values that there is a nil probability of measuring for x and p values which differ *significantly* from the "peak" values, $\langle\hat{X}\rangle_t$ and $\langle\hat{P}\rangle_t$. In such a case, we would be quite justified in saying that the particle at time t is at the point $x(t) \equiv \langle X\rangle_t$ and is moving with momentum $p(t) \equiv \langle P\rangle_t$. This, of course, is just requirement (ii).

The preceding observations have not "proved" that quantum mechanics satisfies requirements (i) and (ii), but they do illustrate fairly well what the general situation is: The observable operators which seem to be required to account for the experimentally observed behavior of real microscopic systems, are found to be such that the universal constant $\hbar$ controls both the spacing between adjacent eigenvalues,

$$A_{n+1} - A_n \simeq \hbar$$

as well as the amount by which two incompatible observables "miss" being compatible,

$$\hat{A}\hat{B} - \hat{B}\hat{A} \simeq \hbar, \qquad \Delta\hat{A} \cdot \Delta\hat{B} \gtrsim \hbar$$

Thus it may be said that the nonclassical features of quantum mechanics owe their existance to the fact that $\hbar$ is *finite* and not zero.† But since from a *macroscopic* point of view $\hbar$ *appears to be*

†The reader who has had some contact with Special Relativity will notice an intriguing analogy here: Relativistic effects are pronounced only when one deals with velocities that are on the order of the velocity of light, c; thus it may be said that the unique features of relativity owe their existence to the fact that c is *finite* and not infinite.

zero, then here all eigenvalues *appear to be continuous* and all observables *appear to be compatible*—and these are just the requirements (i) and (ii).

We turn finally to consider requirement (iii). We have previously shown that $\langle \hat{H} \rangle_t$ is *always* constant in time [see Exercise 43], so it is obviously true that the expectation value of the *energy* satisfies its classical equation of motion [see Exercise 22]. Let us examine the time-dependence of the expectation values of *position* and *momentum*.

The general equation of motion for the expectation value of any observable was derived in Sec. 4-4b and is given by Eq. (4-34). From this equation, it is clear that

$$\frac{d}{dt}\langle \hat{X} \rangle_t = \frac{i}{\hbar}(\Psi_t, [\hat{H}\hat{X} - \hat{X}\hat{H}]\,\Psi_t) \tag{4-71a}$$

and

$$\frac{d}{dt}\langle \hat{P} \rangle_t = \frac{i}{\hbar}(\Psi_t, [\hat{H}\hat{P} - \hat{P}\hat{H}]\,\Psi_t) \tag{4-71b}$$

Now the right-hand sides of these equations can be evaluated explicitly by making use of the expression for $\hat{H}$ in Eq. (4-49a):

Exercise 62. Prove that

$$\hat{H}\hat{X} - \hat{X}\hat{H} = -i\hbar\frac{1}{m}\hat{P} \tag{4-72a}$$

and

$$\hat{H}\hat{P} - \hat{P}\hat{H} = -i\hbar F(\hat{X}) \tag{4-72b}$$

where, in the last equation, the function $F(x)$ is defined by

$$F(x) \equiv -\frac{d}{dx}V(x)$$

so that, by Eqs. (3-2) and (4-8), $F(\hat{X})$ is the *force* operator. [*Hint:* First show that $\hat{X}$ commutes with $V(\hat{X})$ while $\hat{P}$ commutes with $\hat{P}^2/2m$, so that

$$\hat{H}\hat{X} - \hat{X}\hat{H} = \frac{1}{2m}(\hat{P}^2\hat{X} - \hat{X}\hat{P}^2)$$

and

$$\hat{H}\hat{P} - \hat{P}\hat{H} = V(\hat{X})\hat{P} - \hat{P}V(\hat{X})$$

The right-hand side of the first relation is most easily reduced by applying the operator equation (4-68); the right-hand side of the second relation may be reduced by inserting $V(\hat{X}) = V(x)$

and $\hat{P} = -i\hbar(d/dx)$, and then calculating the effect of the resulting operator on some arbitrary function $\phi(x)$.]

When Eqs. (4-72) are inserted into Eqs. (4-71), we immediately obtain the results

$$\frac{d}{dt}\langle\hat{X}\rangle_t = \frac{1}{m}\langle\hat{P}\rangle_t \tag{4-73a}$$

$$\frac{d}{dt}\langle\hat{P}\rangle_t = \langle F(\hat{X})\rangle_t \tag{4-73b}$$

Equations (4-73) are known as *Ehrenfest's equations*. It must be emphasized that these equations are quite general and involve *no* approximations. The first Ehrenfest equation says that $\langle\hat{P}\rangle_t$ and $d\langle\hat{X}\rangle_t/dt$ are related in precisely the same way as the classical momentum p and velocity dx/dt [see Eq. (3-3b)]. The second equation, however, is a bit subtle; by means of the first equation, we can write it as

$$\frac{d}{dt}\left(m\frac{d}{dt}\langle\hat{X}\rangle_t\right) = \langle F(\hat{X})\rangle_t$$

or

$$\frac{d^2}{dt^2}\langle\hat{X}\rangle_t = \frac{1}{m}\langle F(\hat{X})\rangle_t \tag{4-74}$$

Now this equation is *almost* identical to Newton's second law, Eq. (3-3a):

$$\frac{d^2x}{dt^2} = \frac{1}{m}F(x) \qquad [3\text{-}3a]$$

It would be *exactly* identical to Newton's second law *if and only if*

$$\langle F(\hat{X})\rangle_t = F(\langle\hat{X}\rangle_t) \tag{4-75a}$$

for in this case Eq. (4-74) would read

$$\frac{d^2}{dt^2}\langle\hat{X}\rangle_t = \frac{1}{m}F(\langle\hat{X}\rangle_t) \tag{4-75b}$$

which is the same as Newton's second law provided we identify $\langle\hat{X}\rangle_t$ with $x(t)$. In other words, *if* Eq. (4-75a) holds, then $\langle\hat{X}\rangle_t$ will evolve with time in exactly the same way as the position function $x(t)$ does in classical mechanics. Since we have also shown that $\langle\hat{P}\rangle_t$ is related to $\langle\hat{X}\rangle_t$ in the same way that $p(t)$ is related to $x(t)$, then we could conclude that quantum mechanics "corresponds to" classical mechanics in the sense that the time evolutions of $\langle \hat{X} \rangle_t$ and $\langle P \rangle_t$ coincide with the time evolutions of $x(t)$ and $p(t)$, respectively. We repeat, though, that this conclusion hinges on the validity of Eq. (4-75a).

Exercise 63. Show that Eq. (4-75a) *does* hold for the special cases $F(x) = 0$, $F(x) = k$, $F(x) = kx$, where k is any real number; thus conclude that, for these three cases, $\langle \hat{X} \rangle_t$ and $\langle \hat{P} \rangle_t$ obey the usual classical equations of motion. On the other hand, show that Eq. (4-75a) is *not* and identity for the case $F(x) = x^2$.

Now, except for the three cases noted in the above exercise, Eq. (4-75a) is *not* generally valid. The essential reason for this is the same as the reason why $\langle \hat{X}^2 \rangle_t$ is not generally equal to $\langle \hat{X} \rangle_t^2$—namely, the "width" of the position distribution curve

$$\Delta \hat{X}_t = \sqrt{\langle \hat{X}^2 \rangle_t - \langle \hat{X} \rangle_t^2}$$

is *not* always zero [see Fig. 4]. Thus, for state vectors which have a significant dispersion in position, Eq. (4-75b) is not valid for arbitrary force fields $F(x)$, and we must be content with Eq. (4-74).

Suppose, however, we pass to the *macroscopic limit*; here, according to requirement (ii), the width of the position distribution curve will be so small that a series of repeated measurements of the position will yield x-values which are all virtually indistinguishable from the value $\langle \hat{X} \rangle_t$. In such a case, it clearly makes no difference whether we evaluate the average of the $F(x)$-values—i.e., calculate $\langle F(\hat{X}) \rangle_t$—or evaluate F for the average of the x-values—i.e., calculate $F(\langle \hat{X} \rangle_t)$—since we will obtain the same result either way. Therefore, in the *macroscopic limit* Eqs. (4-75) *do* hold for arbitrary force functions $F(x)$, so that, by the foregoing arguments, requirement (iii) is indeed satisfied for the position and momentum observables.

The general relationship which exists between classical mechanics and quantum mechanics is usually referred to as the *Correspondence Principle*. Like so many other things in quantum mechanics, the Correspondence Principle is a very deep and many faceted subject, and we certainly have not exhausted it here. Much of the original thinking on this matter was done by Niels Bohr.

The fundamental connection between the time evolutions of the classical and quantum states was evidently established via the Ehrenfest equations. The fact that these equations fell out of our formalism so simply is certainly a satisfying result, and one which the reader may have found rather surprising. However, it should be noted that we really built this result into quantum mechanics in Postulate 6, when we in effect *used classical mechanics* to tell us what to write down for the Hamiltonian operator $\hat{H}$. In a sense, then, we have *not* completely relegated classical mechanics to the status of a mere "special case" of quantum mechanics. For this reason, it is not altogether clear just what the precise logical relationship is between classical and quantum mechanics. We shall terminate our discussion of this matter with the broad observation that *we ourselves*, as the ultimate observers of any physical system, are essentially classical objects, in that our senses can directly perceive only *macroscopic phenomena* (e.g., dial readings, instrument settings, etc.). This fact probably places severe restrictions not only on what things we can perceive about a *microscopic* system, but also on how we interpret what we perceive. As a result, it is highly questionable if we shall ever be able to discard completely the "crutch" of classical mechanics on the microscopic level.

4-5d A Problem

For a given physical system, it is always important to find the eigenvectors $\{\eta_n(x)\}$ and eigenvalues $\{E_n\}$ of the Hamiltonian operator $\hat{\mathrm{H}}$. There are two reasons for this. First, the numbers E_1, $E_2, \ldots$ provide us with the *allowed energy levels* of the system, and these are always of great practical use in describing the system. Secondly, a knowledge of the functions $\{\eta_n(x)\}$ and the numbers $\{E_n\}$ permit us to write down at once the *solution to the time-evolution problem*; thus, from the given initial state vector $\Psi_0(x)$, we merely calculate the set of complex numbers

$$(\eta_n, \Psi_0) \equiv \int_{-\infty}^{\infty} \eta_n^*(x)\Psi_0(x)dx \qquad n = 1, 2, \ldots \tag{4-76a}$$

which determines the expansion of $\Psi_0(x)$ in the energy eigenbasis,

$$\Psi_0(x) = \sum_{n=1}^{\infty} (\eta_n, \Psi_0)\eta_n(x) \tag{4-76b}$$

and we then have at once an expression for the state vector $\Psi_t(x)$ at any time $t > 0$ [see Eq. (4-42)]:

$$\Psi_t(x) = \sum_{n=1}^{\infty} (\eta_n, \Psi_0)e^{-iE_nt/\hbar}\, \eta_n(x) \tag{4-76c}$$

For a quantum system which has a classical analogue, it is clear that the finding of the energy eigenvectors and eigenvalues is equivalent to solving the time-independent Schrödinger equation for the system's potential function $V(x)$ [see Eq. (4-50)]:

$$-\frac{\hbar^2}{2m}\frac{d^2}{dx^2}\eta_n(x) + V(x)\eta_n(x) = E_n\eta_n(x) \tag{4-77}$$

It would therefore seem a logical next step in our discussion of quantum mechanics to undertake a detailed examination of the general properties of this equation, and to obtain the specific solutions for various "physically important" potential functions. However, this usually turns out to be a complicated and laborious enterprise: Exact solutions to the Schrödinger equation can be found for only a few simple forms for $V(x)$, and even these usually require considerable mathematical expertise. For more complicated potentials, it is necessary to resort to various *approximation techniques*, of which there are a wide variety with varying conditions of applicability. Some of these approximation techniques are quite involved, and almost constitute separate disciplines in their own right.

For these reasons, we shall not delve into the broad problem of solving, either exactly or approximately, the Schrödinger equation for various physical systems. The serious student of physics will find this problem treated exhaustively in existing textbooks on quantum mechanics; indeed, the solution of the Schrödinger equation is usually the major concern of the standard textbooks. In a sense, the aim of this book has not been to *solve* the Schrödinger equation, but rather to place it in the context of a broad (if somewhat simplified) theoretical framework. It is hoped that this will provide the student with an over-all *perspective* of quantum mechanics before he becomes immersed in the complexities of its many applications.

Nevertheless, we cannot in good conscience refrain from working out at least *one* "quantum mechanics problem"—if only to demonstrate that the rather abstract formalism which we have developed in these pages can indeed by applied to a concrete situation. To this end, we shall consider the relatively simple system of a particle of mass m in a one-dimensional "infinite square well," which is the name given by physicists to the potential field defined by

$$V(x) = \begin{cases} 0 & \text{for } |x| \leq L/2 \\ \infty & \text{for } |x| > L/2 \end{cases} \tag{4-78}$$

From the standpoint of classical mechanics, the motion of a particle in this potential field can be understood as follows: Inside the well, $-L/2 < x < L/2$, the particle experiences no force since $F \equiv -dV/dx \equiv 0$, so it moves with a constant momentum, p_0. When the particle strikes the "impenetrable wall" at $x = +L/2$, it experiences an infinite force in the negative x-direction,

$$F(+L/2) = -\left.\frac{dV}{dx}\right|_{L/2} = -\infty$$

which, however, acts for only an infinitesimal time interval; the effect of this impulse is simply to turn the particle around so that its momentum becomes $-p_0$, for this is the only way for the energy of the particle to be conserved. Similarly, when the particle strikes the back wall at a time $L/v_0 = Lm/p_0$ later, its direction of motion is again reversed. Clearly, if the initial state $[x_0, p_0]$ is specified, then it is possible to find the state $[x(t), p(t)]$ at any later time t. Essentially, though, the particle just bounces back and forth between the walls of the well with a period of $2Lm/p_0$, its momentum assuming only the two values $+p_0$ and $-p_0$. The energy of the particle is just

$$E = p_0^2/2m \tag{4-79}$$

and can clearly assume *any* value greater than or equal to zero, depending entirely upon the initial value p_0.

To treat this problem from the standpoint of quantum mechanics, we must first solve the time-independent Schrödinger equation (4-77) for the potential function given in Eq. (4-78). Owing to the way in which this potential function is defined, the Schrödinger equation takes different forms for the two regions $|x| > L/2$ and $|x| \leqq L/2$:

$$|x| > \frac{L}{2}: \quad -\frac{\hbar^2}{2m}\eta_n''(x) + \infty \cdot \eta_n(x) = E_n\eta_n(x) \tag{4-80a}$$

$$|x| \leqq \frac{L}{2}: \quad -\frac{\hbar^2}{2m}\eta_n''(x) = E_n\eta_n(x) \tag{4-80b}$$

The function $\eta_n(x)$ will thus have two pieces, one for "outside" the well, which satisfies Eq. (4-80a), and one for "inside" the well, which satisfies Eq. (4-80b). Clearly, the only solutions to the outside equation are

$$\eta_n(x) \equiv 0 \quad \text{for } |x| > \frac{L}{2} \tag{4-81}$$

Therefore we need consider only the "inside" equation, (4-80b). Now, it can be shown that the infinite jump-discontinuities in $V(x)$ at $x = -L/2$ and $x = +L/2$ render the solutions to the Schrödinger equation *nondifferentiable* but *still continuous* at these two points. Thus, the inside and outside pieces of $\eta_n(x)$ must join at $x = \pm L/2$, but will generally do so with a "kink." In view of Eq. (4-81), the requirement that the inside solutions must connect with the outside solutions implies that these inside solutions must satisfy the *boundary conditions*

$$\eta_n(-L/2) = \eta_n(+L/2) = 0 \tag{4-82}$$

Our problem, then, is to find those functions $\eta_n(x)$ and numbers E_n which satisfy the differential equation (4-80b) and the boundary conditions (4-82). In accordance with Eq. (4-2a), we may expect the functions $\{\eta_n(x)\}$ to form an orthonormal set:

$$(\eta_m, \eta_n) \equiv \int_{-L/2}^{L/2} \eta_m^*(x)\eta_n(x)\,dx = 0, \qquad m \neq n \tag{4-83a}$$

$$(\eta_n, \eta_n) \equiv \int_{-L/2}^{L/2} |\eta_n(x)|^2 \; dx = 1 \tag{4-83b}$$

Exercise 64.

(a) Show that the two functions $A_n \cos k_n x$ and $B_n \sin k_n x$ satisfy Eq. (4-80b), provided $k_n \equiv \sqrt{2mE_n}/\hbar$.

(b) Show that the boundary conditions (4-82) permit solutions of the form $A_n \cos k_n x$ only if $k_n L/2 = n\pi/2$, where $n = 1, 3, 5, \ldots$. Show that solutions of the form $B_n \sin k_n x$ are admissible only if $k_n L/2 = n\pi/2$, where $n = 2, 4, 6, \ldots$.

(c) On the basis of parts (a) and (b), show that the energy eigenvalues for the infinite square well are

$$E_n = \frac{\pi^2 \hbar^2}{2mL^2} n^2 \qquad n = 1, 2, 3, \ldots \tag{4-84}$$

and the corresponding energy eigenvectors are

$$\eta_n(x) = \begin{cases} \sqrt{\dfrac{2}{L}} \cos \dfrac{n\pi}{L} x & n = 1,3,5,\ldots \quad (4\text{-}85a) \\ \sqrt{\dfrac{2}{L}} \sin \dfrac{n\pi}{L} x & n = 2,4,6,\ldots \quad (4\text{-}85b) \end{cases}$$

where the constants A_n and B_n have been chosen in such a way that Eq. (4-83b) is satisfied. Verify that the functions in Eqs. (4-85) satisfy Eqs. (4-83a). *Hint*: For checking the orthonormality conditions, the following identities will be helpful:

$$\int_{-a/2}^{a/2} \sin\frac{n\pi x}{a} \cdot \sin\frac{m\pi x}{a}\, dx = \int_{-a/2}^{a/2} \cos\frac{n\pi x}{a} \cdot \cos\frac{m\pi x}{a}\, dx$$

$$= \begin{cases} 0 & \text{for } m \neq n \\ a/2 & \text{for } m = n \end{cases}$$

$$\int_{-a/2}^{a/2} \sin\frac{n\pi x}{a} \cdot \cos\frac{m\pi x}{a}\, dx = 0$$

(d) Show that the relative spacing of the energy levels around the value E_n is

$$\frac{E_{n+1} - E_n}{E_n} = \frac{2n+1}{n^2} \simeq \frac{2}{n}$$

(e) For a one gram particle moving with velocity 1 cm/sec in a well of length 1 cm, we expect the classical formula for the energy to be correct. Calculate the quantum number n for this case. Would the nearby relative spacings make the energy levels appear discrete or continuous? Would the minimum energy level E_1 be significantly different from zero on this macroscopic scale?

(f) An electron inside a medium weight atom can have energies on the order of 1 kev $= 1.6 \times 10^{-9}$ erg. Calculate the quantum number n for an electron ($m \simeq 10^{-27}$ gram) with an approximate energy of 1 kev, inside a 1 Angstrom well ($L = 10^{-8}$ cm). Would the nearby relative spacings make the energy levels appear discrete or continuous? Would the minimum energy level E_1 be significantly different from zero on this microscopic scale?

(g) Of the three observables, position, momentum, and total energy, if the particle can be said to "have a value" for one, can it be said to "have a value" for either of the other two? [*Hint*: Contrast the forms of the eigenbases $\{\delta_{x_0}(x)\}$, $\{\theta_{p_0}(x)\}$ and $\{\eta_n(x)\}$.]

(h) Show that the uncertainty in the momentum of any mass particle in an infinite square well can never be smaller than $\hbar/2L$.

Exercise 65. Suppose a measurement of the energy at time $t = 0$ yields the number E_n.

(a) Write down the expression for $\Psi_t(x)$ for $t > 0$. What is the position probability density function at any time t? Sketch rough graphs of the position probability density function for the three cases $n = 1$, 2 and 3, and notice that, for a given value of n, there are places in the well where the particle can never be found.

(b) Calculate $\langle\hat{X}\rangle_t$ and $\langle\hat{P}\rangle_t$. Discuss your answers from the standpoint of the concept of the "stationary state." Describe the shape of the energy distribution curve at time t.

(c) Show that when the particle "has energy E_n" then the uncertainties in position and momentum are

$$\Delta\hat{X}_t = \frac{L}{2\sqrt{3}}\sqrt{1 - \frac{6}{\pi^2 n^2}} \quad \text{and} \quad \Delta\hat{P}_t = \frac{\hbar\pi n}{L} = \sqrt{2mE_n}$$

Is the position-momentum uncertainty relation satisfied? Is the time-energy uncertainty relation satisfied?

Exercise 66. Suppose the initial state vector is

$$\Psi_0(x) = \frac{\sqrt{3}}{2}\eta_1(x) + \frac{1}{2}\eta_2(x)$$

(a) What is $\Psi_t(x)$? If the energy of the system is measured at time t, what values can be obtained, and what are their respective probabilities? Using Eqs. (4-11) and (4-12), show that

$$\langle\hat{H}\rangle_t = \left[\frac{\pi^2\hbar^2}{2mL^2}\right]\frac{7}{4} \quad \text{and} \quad \Delta\hat{H}_t = \left[\frac{\pi^2\hbar^2}{2mL^2}\right]\frac{3\sqrt{3}}{4}$$

Prove that the evolution time of any observable cannot be less than $(4mL^2/3\sqrt{3}\pi^2\hbar)$.

(b) Show that the position probability density and the position probability current are given respectively by

$$|\Psi_t(x)|^2 = \frac{3}{4}\eta_1^2(x) + \frac{1}{4}\eta_2^2(x) + \frac{\sqrt{3}}{2}\eta_1(x)\eta_2(x)\cos\left[\frac{E_2 - E_1}{\hbar}t\right]$$

$$S(x,t) = \frac{\hbar}{m}\frac{\sqrt{3}}{4}\left(\eta_1'(x)\eta_2(x) - \eta_1(x)\eta_2'(x)\right)\sin\left[\frac{E_2 - E_1}{\hbar}\right]t$$

Show that the period of oscillation of the density and current are on the order of the evolution time minimum as estimated in part (a). Evaluate the position probability density and current at the points $x = 0$, $x = L/4$, $x = L/3$, $x = L/2$.

(c) Show that the expectation value of the position at time t is

$$\langle \hat{X} \rangle_t = \frac{\sqrt{3}}{2} \left\{ \int_{-L/2}^{L/2} x\eta_1(x)\eta_2(x)\,dx \right\} \cos\left[\frac{E_2 - E_1}{\hbar} t\right]$$

$$= \left(\frac{16\sqrt{3}}{9\pi^2}\right) \frac{L}{2} \cos\left[\frac{3\pi^2\hbar}{2mL^2}\right] t$$

Notice that $\langle \hat{X} \rangle_t$ in this case is *sinusoidal* in time with amplitude $\cong 0.3(L/2)$; by contrast, the classical position function $x(t)$ has a *sawtooth* shape when plotted against t, with amplitude $L/2$.

(d) From either Eq. (4-58a) or (more easily) Eq. (4-73a), show that the expectation value of the momentum at time t is

$$\langle \hat{P} \rangle_t = \frac{-4}{\sqrt{3}} \frac{\hbar}{L} \sin\left[\frac{3\pi^2\hbar}{2mL^2} t\right] = \left(\frac{-8}{\sqrt{21}\pi}\right) \sqrt{2m\langle \hat{H} \rangle} \sin\left[\frac{3\pi^2\hbar}{2mL^2} t\right]$$

where, in the last step, we have made use of the expression for $\langle H \rangle$ in part (a). Notice that $\langle \hat{P} \rangle_t$ in this case is *sinusoidal* in time with amplitude $\cong 0.56\sqrt{2m\langle \hat{H} \rangle}$; by contrast, the classical momentum function $p(t)$ has a *square-wave* shape when plotted against t, with amplitude $\sqrt{2mE}$.

4-6 EXTENSIONS OF THE THEORY

At various stages in our presentation of the theory of quantum mechanics we imposed certain simplifying restrictions so that the main ideas would not be obscured by considerations of slightly lesser importance. It seems appropriate to conclude our development by discussing very briefly how the *removal* of some of these restrictions will affect our simplified picture of quantum mechanics. It must be emphasized that the following discussion is not meant to be as detailed as that in the previous sections. Our purpose now is merely to get a rough idea of what is involved in obtaining a more general theory, and to thereby establish a bridge to the reader's future studies in quantum mechanics.

4-6a Systems with More Than One Degree of Freedom

The first major restriction which we imposed was that the physical system have only one degree of freedom. We labeled this degree of freedom by the variable or "coordinate" x. Now, most systems of interest in physics have *more* than one degree of freedom. For example, a particle in real space will have three degrees of freedom corresponding to its position (i.e., x,y,z or r,θ,ϕ), and it may also have additional degrees of freedom due to its orientation or to some internal structure; again, a system consisting of two particles with no other attributes will generally have six degrees of freedom, which can be labelled by the six coordinates $x_1, y_1, z_1, x_2, y_2, z_2$.

It is not difficult, at least from a formal standpoint, to adapt our treatment of a system with one degree of freedom to a system with n degrees of freedom. To do this, we first associate with each degree of freedom a so-called "generalized coordinate" q_i. The set of generalized coordinates $q_1, q_2, \ldots, q_n$ may consist of cartesian coordinates, angles, and in general any group of variables which, when taken together, specify the "configuration" of the system in the same way that x specifies the configuration of a particle in one dimension. Having done this, we then set up a Hilbert space for the system. The *vectors* in this Hilbert space are all those complex functions ψ of n real variables $q_1, q_2, \ldots, q_n$ which satisfy [see Eq. (2-35)]

$$\int - \int |\psi(q_1, \ldots, q_n)|^2 \, dq_1 \cdots dq_n < \infty \tag{4-86}$$

Here, the integrations are to be carried out over the full ranges of the various coordinates q_i. From this point on, the development of the properties of the Hilbert space is *identical* to that given in Secs. 2-3 and 2-4, except that the single variable x is everywhere replaced by the n variables $q_1, \ldots, q_n$. Thus, for example, the definition of the inner product in Eq. (2-32) becomes

$$(\psi_1, \psi_2) \equiv \int - \int \psi_1^*(q_1, \ldots, q_n)\psi_2(q_1, \ldots, q_n) dq_1 \cdots dq_n \tag{4-87}$$

The possible state vectors of the system are, as before, all the normed vectors in $\mathcal{H}$; we write the state vector at time t as $\Psi_t(q_1, \ldots, q_n)$ or $\Psi(q_1, \ldots, q_n; t)$.

Now, in advanced treatments of classical mechanics, there is developed a well-defined procedure for associating with any "generalized coordinate" q_i a "generalized momentum" p_i. We shall not go into this procedure here, but we might just mention by way of example that, if $q_i = x$ then p_i turns out to be $p_x = mv_x$, and if $q_i = \theta$ then p_i turns out to be the angular momentum associated with the rotation of θ. In any case, the two sets of variables $\{q_i\}$ and $\{p_i\}$ form the basic *observables* of both the classical description and the quantum description. In direct analogy with Eqs. (4-46) and (4-47), we postulate that the quantum operators corresponding to q_i and p_i are

$$\hat{Q}_i = q_i \qquad \text{and} \qquad \hat{P}_i = -i\hbar \frac{\partial}{\partial q_i} \tag{4-88}$$

and, moreover, that the operator corresponding to any observable $f(q_1, p_1, \ldots, q_n, p_n)$ is [see Eq. (4-48)]

$$f\left(q_1, -i\hbar\frac{\partial}{\partial q_1}, \ldots, q_n, -i\hbar\frac{\partial}{\partial q_n}\right) \tag{4-89}$$

As in Eqs. (4-68) and (4-69), the above definitions for $\hat{Q}_i$ and $\hat{P}_i$ imply that

$$\hat{Q}_i\hat{P}_i - \hat{P}_i\hat{Q}_i = i\hbar \tag{4-90}$$

so that, by the Heisenberg Uncertainty Principle,

$$\Delta\hat{Q}_i \cdot \Delta\hat{P}_i \geqq \frac{\hbar}{2} \tag{4-91}$$

Although $\hat{Q}_i$ and $\hat{P}_i$ do *not* commute with each other, they *do* commute with all the rest of the $\hat{Q}_j$'s and $\hat{P}_j$'s. In particular, this implies by the Compatibility Theorem that the generalized coordinates are all *simultaneously measureable.* By the same sort of arguments used to derive Eq. (4-61), it can be shown that the quantity

$$|\Psi_t(q_1, \ldots, q_n)|^2 \, dq_1 \cdots dq_n \tag{4-92}$$

is the probability that a simultaneous measurement of all the coordinates at time t will yield a value for q_1 in the interval $(q_1, q_1 + dq_1)$, *and* a value for q_2 in the interval $(q_2, q_2 + dq_2)$, . . . , *and* a value for q_n in the interval $(q_n, q_n + dq_n)$.

For the important case of a simple particle in three dimensions, we can write the state vector as $\Psi_t(x,y,z) \equiv \Psi(x,y,z;t)$. Since the classical Hamiltonian function is now

$$H(x,y,z,p_x,p_y,p_z) = \frac{1}{2m}(p_x^2 + p_y^2 + p_z^2) + V(x,y,z)$$

then replacing the variable x by the operator x, the variable p_x by the operator $-i\hbar\,(\partial/\partial x)$, etc., we obtain in analogy with Eqs. (4-50) and (4-51) the time-independent and the time-dependent Schrödinger equations:

$$-\frac{\hbar^2}{2m}\left[\frac{\partial^2}{\partial x^2}+\frac{\partial^2}{\partial y^2}+\frac{\partial^2}{\partial z^2}\right]\eta_n(x,y,z) + V(x,y,z)\eta_n(x,y,z) = E_n\eta_n(x,y,z) \tag{4-93}$$

$$-\frac{\hbar^2}{2m}\left[\frac{\partial^2}{\partial x^2}+\frac{\partial^2}{\partial y^2}+\frac{\partial^2}{\partial z^2}\right]\Psi(x,y,z;t) + V(x,y,z)\Psi(x,y,z;t) = i\hbar\frac{\partial}{\partial t}\Psi(x,y,z;t) \tag{4-94}$$

Thus, for example, if one solves Eq. (4-93) for the Coulomb potential, $V(x,y,z) = \frac{-e^2}{4\pi\epsilon_0}[x^2+y^2+z^2]^{-\frac{1}{2}}$, one obtains for the eigenvalues $\{E_n\}$ just the "Bohr energy levels" for the hydrogen atom. Of course, one *also* obtains, through the associated eigenvectors $\{\eta_n(x)\}$, *much more* information about the hydrogen atom than was available via the old Bohr theory.

When we extend our treatment of a particle from one dimension to three dimensions, we find that another important observable comes into play: In classical mechanics, a particle at the position **r** with momentum **p** has an "angular momentum about the origin"—or, as we shall prefer to say, an *orbital angular momentum*—given by $\ell = \mathbf{r} \times \mathbf{p}$. In component form, this means that

$$\ell_x = yp_z - zp_y, \quad \ell_y = zp_x - xp_z, \quad \ell_z = xp_y - yp_x \tag{4-95}$$

Therefore, according to the generalized form of Postulate 6, the quantum *operator* for the observable ℓ_x is

$$\hat{L}_x = \hat{Y}\hat{P}_z - \hat{Z}\hat{P}_y = -i\hbar\left(y\frac{\partial}{\partial z} - z\frac{\partial}{\partial y}\right) \tag{4-96}$$

Analogous expressions follow for $\hat{L}_y$ and $\hat{L}_z$. Now from these definitions of $\hat{L}_x$, $\hat{L}_y$ and $\hat{L}_z$, along with the commutation relations $\hat{X}\hat{P}_x - \hat{P}_x\hat{X} = i\hbar$, $\hat{Y}\hat{P}_y - \hat{P}_y\hat{Y} = i\hbar$, and $\hat{Z}\hat{P}_z - \hat{P}_z\hat{Z} = i\hbar$, it is fairly straightforward to establish the following commutation relations among the three orbital angular momentum component operators:

$$\left.\begin{aligned} \hat{L}_x\hat{L}_y - \hat{L}_y\hat{L}_x &= i\hbar\hat{L}_z \\ \hat{L}_y\hat{L}_z - \hat{L}_z\hat{L}_y &= i\hbar\hat{L}_x \\ \hat{L}_z\hat{L}_x - \hat{L}_x\hat{L}_z &= i\hbar\hat{L}_y \end{aligned}\right\}. \tag{4-97}$$

Exercise 67. Derive the first of Eqs. (4-97). Note that the other two can be obtained by cyclically permuting the indices.

One then generalizes and calls *any* three operators which satisfy the above commutation relations the "components of an *angular momentum operator.*" The general treatment of angular momentum in quantum mechanics is somewhat involved. However, a great many important physical phenomena find their interpretations in terms of the various properties of angular momentum operators.

4-6b Some Remarks on Continuous Eigenvalues

Our development of the theory of quantum mechanics in the first four sections of this chapter was restricted to observables with *discretely* distributed eigenvalues. However, in Sec. 4-5a we had to abandon this restriction to some extent, since the position and momentum operators of Postulate 6 were found to have *continuously* distributed eigenvalues. In connection with this, we encountered some unusual mathematical difficulties associated with the eigenbasis vectors $\{\delta_{x_0}(x)\}$ and $\{\theta_{p_0}(x)\}$ of these two operators [see the discussion following Exercise 50]. Similar difficulties arise in the *general* treatment of operators with continuously distributed eigenvalues. The handling of these difficulties is somewhat involved, but pivots mainly upon the so-called *Dirac delta function*, $\delta(x - x_0)$, a highly unusual entity which just happens to coincide with the position eigenvector $\delta_{x_0}(x)$.

We remarked in our discussion of $\delta_{x_0}(x)$ in Sec. 4-5a that this function should vanish for $x \neq x_0$, but that $\delta_{x_0}(x_0)$ should be *infinite* in such a way that its product with the infinitesimal dx be *finite*. In fact, the actual definition of the Dirac delta function is such that

$$\delta_{x_0}(x_0)dx \equiv \delta(0)dx = 1$$

More specifically, the Dirac delta function $\delta(x - x_0)$ is *defined* by the following two relations:

$$\begin{cases} \delta(x - x_0) = 0 \quad \text{for } x \neq x_0 & \text{(4-98a)} \\ \displaystyle\int_a^b \delta(x - x_0)dx = 1 \quad \text{for any } a < x_0 < b & \text{(4-98b)} \end{cases}$$

Exercise 68. Using these defining properties of the Dirac delta function, derive the following two properties:

$$\delta(x - x_0) = \delta(x_0 - x) \tag{4-99}$$

$$f(x) = \int_{-\infty}^{\infty} f(x')\delta(x' - x)dx' \tag{4-100}$$

[*Hint:* To establish Eq. (4-100), multiply the arbitrary function value $f(x)$ by $1 = \int \delta(x' - x)dx'$, move $f(x)$ inside the x'-integral, and then given an argument for replacing the $f(x)$ by $f(x')$.]

We recall that, for an observable operator $\hat{A}$ whose eigenvalues and eigenvectors can be labeled by a *discrete* index n,

$$\hat{A}\alpha_n(x) = A_n \alpha_n(x) \tag{4-101a}$$

the eigenvectors $\{\alpha_n(x)\}$ must form an orthonormal basis in the Hilbert space:

$$(\alpha_n, \alpha_{n'}) = \delta_{nn'} \qquad \text{for all } n,n' \tag{4-101b}$$

$$\phi(x) = \sum_n (\alpha_n, \phi)\alpha_n(x) \qquad \text{for all } \phi(x) \text{ in } \mathcal{H} \tag{4-101c}$$

Now for the case in which the eigenvalues and eigenvectors of $\hat{A}$ must be labeled by a *continuous* index ν,

$$\hat{A}\alpha_\nu(x) = A(\nu)\alpha_\nu(x) \tag{4-102a}$$

it is necessary to modify somewhat our definition of an orthonormal basis set. Specifically, Eq. (4-101b) is modified by replacing the Kronecker delta symbol δ_{ab} by the Dirac delta function $\delta(a - b)$,

$$(\alpha_\nu, \alpha_{\nu'}) = \delta(\nu - \nu') \qquad \text{for all } \nu,\nu' \tag{4-102b}$$

and Eq. (4-101c) is modified by replacing the sum over the discrete index n by an integral over the continuous index ν,

$$\phi(x) = \int (\alpha_\nu, \phi)\alpha_\nu(x)d\nu \qquad \text{for all } \phi(x) \text{ in } \mathcal{H} \tag{4-102c}$$

Finally, the rule in Postulate 3, that $|(\alpha_n, \Psi_t)|^2$ gives the probability of measuring the discrete value A_n in the state $\Psi_t(x)$, takes the following form for the case of continuous eigenvalues: *If $\mathfrak{A}$ is measured on the state $\Psi_t(x)$, the probability of obtaining a value between $A(\nu)$ and $A(\nu + d\nu)$ is $|(\alpha_\nu, \Psi_t)|^2 \, d\nu$.*

The mathematical transition from Eq. (4-101c) to Eq. (4-102c) is a rather natural one; however, the transition from Eq. (4-101b) to Eq. (4-102b) is not altogether obvious at first sight. In order to understand why we write $\delta(\nu - \nu')$ in Eq. (4-102b) rather than just $\delta_{\nu\nu'}$, let us make the following calculation: let us write out explicitly the definition of the quantity $(\alpha_{\nu'},\phi)$, and then insert for $\phi(x)$ the expansion in Eq. (4-102c). We have

$$(\alpha_{\nu'},\phi) \equiv \int_{-\infty}^{\infty} \alpha_{\nu'}^{*}(x)\phi(x)\,dx = \int_{-\infty}^{\infty} \alpha_{\nu'}^{*}(x)\left[\int(\alpha_{\nu},\phi)\alpha_{\nu}(x)\,d\nu\right]dx$$

Interchanging the order of the x- and ν-integrations, we obtain

$$(\alpha_{\nu'},\phi) = \int(\alpha_{\nu},\phi)\left[\int_{-\infty}^{\infty} \alpha_{\nu'}^{*}(x)\alpha_{\nu}(x)\,dx\right]d\nu \equiv \int(\alpha_{\nu},\phi)\,(\alpha_{\nu'},\alpha_{\nu})\,d\nu$$

Now the quantity (α_{ν},ϕ) is a complex number which depends on the index ν, or in other words, (α_{ν},ϕ) is some arbitrary complex function of the real variable ν; writing this function $c(\nu)$, the last equation becomes

$$c(\nu') = \int c(\nu)\,(\alpha_{\nu'},\alpha_{\nu})\,d\nu$$

Clearly, if $(\alpha_{\nu'},\alpha_{\nu}) = \delta_{\nu\nu'}$, then the integral on the right would vanish, thereby rendering the equation incorrect. Indeed, the *only* way for this equation to hold true for any arbitrary function $c(\nu)$ is for $(\alpha_{\nu'},\alpha_{\nu})$ to be the Dirac delta function $\delta(\nu' - \nu)$, in which case the equation is simply an instance of Eq. (4-100). We see then that, if we want Eq. (4-102c) to be valid for all functions $\phi(x)$, we are essentially *forced* to require the functions $\{\alpha_{\nu}(x)\}$ to satisfy Eq. (4-102b). As an example, the position and momentum eigenvectors are supposed to satisfy

$$(\delta_{x_1},\delta_{x_2}) = \delta(x_1 - x_2) \qquad -\infty < x_1,x_2 < \infty \qquad \text{(4-103a)}$$

and

$$(\theta_{p_1},\theta_{p_2}) = \delta(p_1 - p_2) \qquad -\infty < p_1,p_2 < \infty \qquad \text{(4-103b)}$$

Exercise 69. Using the definition of the inner product, along with Eq. (4-100), prove that the position eigenvectors do indeed satisfy the modified orthonormality relation in Eq. (4-103a). [*Hint:* Remember that $\delta_{x_0}(x) \equiv \delta(x - x_0)$ is pure real.]

We cannot discuss here all the ramifications of these generalizations for continuous eigenvalues. In particular, Eq. (4-103b), and the precise relationship between the functions $\delta_{x_0}(x)$ and $\theta_{p_0}(x)$, can be fully appreciated only in the context of an area of mathematics known as Fourier analysis. However, there is one important consequence of all this that perhaps should be brought out. It will be recalled that any $\mathcal{H}$-vector $\phi(x)$ may be said to have "components"

$$(\alpha_n, \phi) \qquad n = 1, 2, \ldots \tag{4-104a}$$

relative to a given or orthonormal basis set $\{\alpha_n(x)\}$, just as any $\mathcal{E}_3$-vector **v** has components

$$\mathbf{e}_n \cdot \mathbf{v} \qquad n = 1, 2, \text{and } 3 \tag{4-104b}$$

relative to a given orthonormal basis set $\mathbf{e}_1$, $\mathbf{e}_2$, $\mathbf{e}_3$ [see Eqs. (2-29c) and (2-39c)]. Thus any vector in $\mathcal{H}$ can be "represented" relative to a given orthonormal basis by an ∞-*tuple of complex numbers*, in the same sense that any vector in $\mathcal{E}_3$ can be represented relative to a given orthonormal basis by a triplet of real numbers. Evidently, the components of $\phi(x)$ relative to the continuous eigenbasis $\{\alpha_\nu(x)\}$ will be labelled by the continuous index ν, (α_ν, ϕ). In particular, let us calculate the components of $\phi(x)$ relative to the *position* eigenbasis, $\{\delta_\nu(x)\}$.

Exercise 70. Using the definition of the inner product, along with Eq. (4-100), prove that

$$(\delta_\nu, \phi) = \phi(\nu) \qquad -\infty < \nu < \infty \tag{4-105}$$

Therefore, relative to the eigenbasis of the *position* operator $\hat{X}$, a given *vector* $\phi(x)$—i.e., a given *function* ϕ of x—has *components* which are just the set of all the *values* of this function $c_\nu \equiv \phi(\nu)$. In this sense we may say that any vector $\phi(x)$ in $\mathcal{H}$ "represents itself" with respect to the *position* eigenbasis. Indeed, the expansion of $\phi(x)$ in the position eigenbasis $\{\delta_\nu(x)\}$ is, according to Eqs. (4-102c) and (4-105),

$$\phi(x) = \int_{-\infty}^{\infty} (\delta_\nu, \phi)\delta_\nu(x)\,d\nu = \int_{-\infty}^{\infty} \phi(\nu)\delta(x - \nu)\,d\nu \tag{4-106}$$

which is evidently nothing more than Eq. (4-100).

Equation (4-105) provides us with an interesting interpretation of the definition of the inner product, Eq. (2-32), as the following exercise demonstrates.

Exercise 71. In Exercise 13 we derived Eq. (2-40a), which gives the inner product of two $\mathcal{H}$-vectors in terms of their *components* relative to a given orthonormal basis $\{\epsilon_i(x)\}$. If we take for $\{\epsilon_i(x)\}$ the position eigenbasis $\{\delta_\nu(x)\}$, and if we replace the sum over the discrete index n by an integral over the continuous index ν, show that Eq. (2-40a) takes the form of our original *definition* of the inner product in Eq. (2-32). [*Hint*: In Eq. (2-40a), recall that $c_i = (\epsilon_i, \psi)$ and $d_i = (\epsilon_i, \phi)$.]

Finally, we should note that, in accordance with the previously mentioned interpretation of $|(\alpha_\nu,\Psi_t)|^2\, d\nu$ as being a measurement probability, the quantity $|(\delta_\nu,\Psi_t)|^2\, d\nu$ is evidently supposed to represent the probability of measuring for the *position* a value between ν and $\nu+d\nu$. But, by Eq. (4-105), we see that

$$|(\delta_\nu,\Psi_t)|^2\, d\nu = |\Psi_t(\nu)|^2\, d\nu$$

so we have recovered the "Born interpretation" of the state vector in Eq. (4-61).

As the reader may have sensed, our definition and use of the symbol $\delta(x - x_0)$ plays "fast and loose" with the laws of calculus. A rigorous treatment of the Dirac delta function requires a rather lengthy sojourn into an area of mathematics known as Distribution Theory; our treatment here is illustrative of the more "intuitive" approach taken by most standard textbooks on quantum mechanics, and the reader is referred to any of these books for a more detailed analysis of the Dirac delta function.

4-6c The Problem of Degeneracy

We wish now to discuss the effects of removing the restriction in Eq. (4-9) that the eigenvalues of an observable operator be unequal or *nondegenerate*. To this end, let us examine the simple case in which a certain eigenvalue of an observable operator $\hat{A}$ is *doubly degenerate;* more specifically, we suppose that the eigenvectors $\alpha_1(x)$ and $\alpha_2(x)$ correspond to the same eigenvalue A_{12}, but that no other eigenvector of $\hat{A}$ has this eigenvalue:

$$A_1 = A_2 = A_{12}, \qquad \text{but} \quad A_i \neq A_{12} \ \text{for } i \geqq 3 \qquad (4\text{-}107)$$

This circumstance will necessitate modifications in both Postulates 3 and 4. As indicated in our discussion just preceding Eq. (4-9), the required modification of Postulate 3 is merely a straightforward application of the "addition rule" for probabilities in Eq. (2-3a):

Postulate 3′. If the eigenvalues of $\hat{A}$ satisfy Eq. (4-107), and if $\mathcal{A}$ is measured on the state $\Psi_t(x)$, then the probability for obtaining the eigenvalue A_{12} is $|(\alpha_1,\Psi_t)|^2 + |(\alpha_2,\Psi_t)|^2$.

Of somewhat more interest is the modified form of Postulate 4, which reads as follows:

Postulate 4′. If the eigenvalues of $\hat{A}$ satisfy Eq. (4-107), and if a measurement of $\mathcal{A}$ on the state $\Psi_t(x)$ yields the eigenvalue A_{12}, then the state vector of the system immediately after the measurement is given by

$$\Psi'_{A_{12}}(x) = \frac{(\alpha_1,\Psi_t)\alpha_1(x) + (\alpha_2,\Psi_t)\alpha_2(x)}{\sqrt{|(\alpha_1,\Psi_t)|^2 + |(\alpha_2,\Psi_t)|^2}} \tag{4-108}$$

This last equation may appear strange at first, but it actually has a very simple interpretation: We can write the state vector immediately *before* the measurement as [see Eq. (4-6a)]

$$\Psi_t(x) = \left\{(\alpha_1,\Psi_t)\alpha_1(x) + (\alpha_2,\Psi_t)\alpha_2(x)\right\} + \sum_{i=3}^{\infty} (\alpha_i,\Psi_t)\alpha_i(x)$$

In view of this, we see that Postulate 4′ merely asserts that a measurement of $\mathfrak{A}$ *with the result* A_{12} essentially "wipes out" that portion of $\Psi_t(x)$ corresponding to eigenvalues *other than* A_{12}, but "passes undistorted" the parts of the state vector which *belong* to the eigenvalue A_{12}. The denominator in the above expression for $\Psi'_{A_{12}}(x)$ is merely to make the vector properly normalized.

Exercise 72. Prove that the vector $\Psi'_{A_{12}}(x)$ has unit norm.

It should be noted that the form of Postulate 4 which we presented in Sec. 4-3b is just a special case of Postulate 4′; for if the eigenvalues of $\hat{A}$ are nondegenerate, then by Postulate 4′, a measurement of $\mathfrak{A}$ with the result A_1 will leave the system in the state

$$\Psi'_{A_1}(x) = \frac{(\alpha_1,\Psi_t)}{|(\alpha_1,\Psi_t)|}\alpha_1(x)$$

But since the complex number multiplying $\alpha_1(x)$ obviously has unit modulus, then $\Psi'_{A_1}(x)$ may be said to coincide with $\alpha_1(x)$ in the sense allowed by Postulate 1. This is just the statement of Postulate 4 in Sec. 4-3b.

In developing the various consequences of these more general forms of the two "measurement postulates," the following simple theorem plays a key role.

Exercise 73. Prove that, if $\alpha_1(x)$ and $\alpha_2(x)$ are eigenvectors of $\hat{A}$ belonging to the *same* eigenvalue, then any linear combination of these two vectors, $c_1\alpha_1(x) + c_2\alpha_2(x)$, is *also* an eigenvector of $\hat{A}$ belonging to this eigenvalue. [*Hint:* Using the linearity of $\hat{A}$, examine the effect of $\hat{A}$ acting on the linear combination.]

Since $\Psi'_{A_{12}}(x)$ in Postulate 4′ is evidently a linear combination of $\alpha_1(x)$ and $\alpha_2(x)$, then one consequence of the foregoing theorem is that the following statement is *generally valid*: A measurement of

$\mathcal{A}$ forces the state vector of the system into *an* eigenvector of $\hat{A}$ belonging to the eigenvalue obtained in the measurement. If the eigenvalue measured is nondegenerate, then there is only one corresponding eigenvector, and it is with this eigenvector that the state vector will coincide after the measurement. However, if the eigenvalue measured is degenerate, then there are infinitely many corresponding eigenvectors (i.e., all possible linear combinations of the "standard" eigenvectors), and it is necessary to know the state vector of the system *before* the measurement in order to determine unambiguously the state vector *after* the measurement.

A second consequence of Postulates 3′ and 4′, and the theorem of Exercise 73, is that the Compatibility Theorem which we presented in Sec. 4-3c remains valid as stated. However, the possibility of degenerate eigenvalues confers upon the Compatibility Theorem a new importance, which we shall now attempt to explain:

Suppose again that the eigenvectors $\alpha_1(x)$ and $\alpha_2(x)$ of the observable operator $\hat{A}$ correspond to the same eigenvalue A_{12}. Then according to Exercise 73, the two vectors $\tilde{\alpha}_1(x)$ and $\tilde{\alpha}_2(x)$ defined by

$$\tilde{\alpha}_1(x) = c_1\alpha_1(x) + c_2\alpha_2(x)$$

and

$$\tilde{\alpha}_2(x) = c_3\alpha_1(x) + c_4\alpha_2(x)$$

where c_1, c_2, c_3 and c_4 are any complex numbers, are also eigenvectors of $\hat{A}$ belonging to the eigenvalue A_{12}; in addition, it is clear that these two eigenvectors are orthogonal to all the other eigenvectors $\alpha_3(x)$, $\alpha_4(x)$, ... Now, it is possible to choose the c_i-numbers in such a way that $\tilde{\alpha}_1(x)$ and $\tilde{\alpha}_2(x)$ have unit norms and are orthogonal to each other; in fact, there are infinitely many ways of doing this. We can see this most easily if we take as an analogy a three-dimensional vector space in which the basis vectors $\mathbf{e}_1$ and $\mathbf{e}_2$ are respectively identified with the eigenvectors $\alpha_1(x)$ and $\alpha_2(x)$, while the basis vector $\mathbf{e}_3$ represents all the other eigenvectors $\alpha_3(x)$, $\alpha_4(x)$, . . . , all of which are orthogonal to the first two vectors. In this analogy, $\mathbf{e}_1$ and $\mathbf{e}_2$ are eigenvectors belonging to the eigenvalue A_{12}; but, according to Exercise 73, any vector in the plane formed by $\mathbf{e}_1$ and $\mathbf{e}_2$ is also an eigenvector of $\hat{A}$ with this same eigenvalue. This being the case, we need not tie ourselves down to the orthonormal pair $\mathbf{e}_1$ and $\mathbf{e}_2$, but we may evidently use any pair $\tilde{\mathbf{e}}_1$ and $\tilde{\mathbf{e}}_2$ which differs from $\mathbf{e}_1$ and $\mathbf{e}_2$ by a simple rotation about the $\mathbf{e}_3$-axis; for such a pair would belong to the same eigenvalue of $\hat{A}$ as $\mathbf{e}_1$ and $\mathbf{e}_2$, and moreover, the vectors $\tilde{\mathbf{e}}_1$, $\tilde{\mathbf{e}}_2$ and $\mathbf{e}_3$ would clearly constitute an orthonormal basis set. In a similar way, we can always replace $\alpha_1(x)$ and $\alpha_2(x)$ with any one of an infinite number of orthonormal pairs

$\tilde{\alpha}_1(x)$ and $\tilde{\alpha}_2(x)$ in the "plane" of $\alpha_1(x)$ and $\alpha_2(x)$; from the point of view of the operator $\hat{A}$, it makes no difference at all whether we choose for its eigenbasis the orthonormal set $\alpha_1(x), \alpha_2(x), \alpha_3(x), \ldots$ or the orthonormal set $\tilde{\alpha}_1(x), \tilde{\alpha}_2(x), \alpha_3(x), \ldots$.

With this point understood, suppose we now introduce a second observable operator $\hat{B}$ which commutes with $\hat{A}$. According to the Compatibility Theorem, $\hat{A}$ and $\hat{B}$ possess a common eigenbasis. However, this is *not* to say that *every* eigenvector of $\hat{A}$ in the "plane" of $\alpha_1(x)$ and $\alpha_2(x)$ is necessarily an eigenvector of $\hat{B}$ as well; all we can conclude from the Compatibility Theorem is that *at least two* eigenvectors in this plane, say $\tilde{\alpha}_1(x)$ and $\tilde{\alpha}_2(x)$, are orthonormal eigenvectors of $\hat{B}$. Now, if it should happen that $\tilde{\alpha}_1(x)$ and $\tilde{\alpha}_2(x)$ do not correspond to the same eigenvalue of $\hat{B}$, then they will in fact be the *only* eigenvectors of $\hat{B}$ in the plane of $\alpha_1(x)$ and $\alpha_2(x)$.† Consequently, the common eigenbasis $\{\phi_n(x)\}$ will necessarily contain $\tilde{\alpha}_1(x)$ and $\tilde{\alpha}_2(x)$ and *not* $\alpha_1(x)$ and $\alpha_2(x)$. In a manner of speaking, the introduction of the compatible observable $\mathcal{B}$ has "resolved" the ambiguity of which of the infinitely many pairs of A_{12}-eigenvectors "ought" to be used; at the same time, of course, it is likely that other ambiguities in either the $\hat{A}$ eigenvectors or the $\hat{B}$ eigenvectors become similarly resolved.

Now, it is clear that the introduction of the compatible observable $\mathcal{B}$ may not completely resolve all degeneracies; that is, there may still be "subspaces" of two or even more dimensions such that all vectors in any one of these subspaces are eigenvectors belonging to the same eigenvalue of $\hat{A}$ *and* the same eigenvalue of $\hat{B}$. In such a case, we must search for a third observable $\mathcal{C}$ whose operator $\hat{C}$ commutes with *both* $\hat{A}$ and $\hat{B}$, and which further reduces the number of degeneracies.‡ We continue in this way until we obtain what is called a *complete set of compatible observables*, $\mathcal{A}, \mathcal{B}, \ldots, \mathfrak{F}$. The corresponding operators, $\hat{A}, \hat{B}, \ldots, \hat{F}$ all commute in pairs, and possess a common eigenbasis $\{\phi_n(x)\}$ which has the following special property: each eigenvector of the set $\{\phi_n(x)\}$ corresponds to a unique set of eigenvalues of the operators $\hat{A}, \hat{B}, \ldots, \hat{F}$. In other words, any two of the eigenvectors $\phi_m(x)$ and $\phi_n(x)$ belong to different eigenvalues of at least one of these operators. This, in fact, is just the defining condition for a complete set of compatible observables. It is a basic

†This follows from the theorem proved in Exercise 18: Two eigenvectors of a Hermitian operator belonging to *unequal* eigenvalues must be orthogonal.

‡We emphasize "both" here, because given that $\hat{A}$ and $\hat{B}$ commute, then the fact that $\hat{C}$ commutes with $\hat{A}$ does *not* guarantee that $\hat{C}$ commutes with $\hat{B}$ as well. As an example, one can show from Eqs. (4-88) that the operator $\hat{L}_x^2 + \hat{L}_y^2 + \hat{L}_z^2$ commutes with both $\hat{L}_x$ and $\hat{L}_y$, even though $\hat{L}_x$ and $\hat{L}_y$ clearly do not commute with each other.

assumption of quantum mechanics that every physical system possesses at least one such set of observables.

The concept of a complete set of compatible observables leads us to another important concept, namely, that of a "maximal measurement." We recall that, for the case of nondegenerate eigenvalues, Postulate 4 tells us that a measurement of a *single* observable is sufficient to "prepare a state"—i.e., to force the system into a state vector which is *known*, even though the state vector just prior to the measurement was *unknown*. However, when degenerate eigenvalues are involved, we cannot make this statement; for, if the measurement should yield a degenerate eigenvalue, then Postulate 4′ clearly implies that the resulting state vector cannot be uniquely determined unless the initial state vector is known. How, then, can we "prepare a state" when degenerate eigenvalues are involved? The answer to this question is simply that we must simultaneously measure *all* the observables of a complete, compatible set. The performance of these simultaneous or successive measurements, which will not "interfere" with one another since the observables are by definition compatible, will necessarily leave the state vector of the system coincident with a vector which is at once an eigenvector of *all* the operators $\hat{A}$, $\hat{B}, \ldots, \hat{F}$—i.e., with one of the simultaneous eigenbasis vectors $\{\phi_n(x)\}$. Since the observables are "complete," we then need only examine the eigenvalues $A_i, B_j, \ldots, F_k$ obtained in these measurements to pinpoint precisely *which one* of the common eigenvectors the state vector of the system has been forced into. The preparation of a state by simultaneously measuring all of the observables of a complete, compatible set is called a *maximal measurement* of the system. It is clear that a single measurement of an observable whose eigenvalues are completely nondegenerate is already a maximal measurement.

The presence of degenerate eigenvalues can be directly observed in a variety of laboratory experiments. For example, the Hamiltonian operator for a number of electrons in the pure Coulomb field of an atomic nucleus is found to commute with the angular momentum operator, but to be highly degenerate; in other words, it is found that many of the common energy/angular momentum eigenbasis vectors will correspond to the *same* eigenvalue of the energy operator, but *different* eigenvalues of the angular momentum operator. If an external magnetic field is impressed upon the atom, the Hamiltonian operator will be altered, because the classical Hamiltonian function will now contain terms describing the interaction energy between the electrons and the external field. Now, under certain conditions, the new Hamiltonian operator will *still* commute with the angular momentum operator, but it will *not* be so highly degenerate. Thus a set of angular momentum eigenvalues which previously corresponded to the *same* energy eigenvalue will now correspond to *different* energy eigenvalues. The result is termed a "splitting of the energy levels," and the effects can be observed and quantitatively accounted for in spectroscopic studies in the laboratory.

Let us summarize now the important modifications to our theory which are necessitated by the presence of degenerate eigenvalues: Postulate 3 is modified, but in a very obvious way. Postulate 4 is modified in such a way as to suggest that an ideal measurement is analogous to sending the state vector of the system through a "filter," which passes undistorted all those components of the state vector which belong to the measured eigenvalue, but which completely cuts out all the other components. The Compatibility Theorem remains intact, but a unique choice for the vectors of the common eigenbasis set $\{\phi_n(x)\}$ requires the specification of a "complete set of compatible observables." In order for the measurement process to yield a state vector which is known even though the state vector just prior to the measurement is not known—i.e., in order to "prepare a state"—it is necessary to make a "maximal measurement"; this is simply a set of simultaneous measurements of all the members of a complete set of compatible observables, and it results in a precisely known state, irrespective of the initial state, simply because each of the common eigenbasis vectors $\{\phi_n(x)\}$ corresponds to a unique set of eigenvalues for these observables.

4-6d Concluding Remarks

In closing, we might recall that the theory of quantum mechanics given here is restricted to systems which are *nonrelativistic*. In other words, we have presented a theory which is valid only for systems which contain velocities that are much less than the velocity of light, c, or equivalently, for systems which contain particles whose energies are much less than their rest-mass energies, mc^2. We remarked in Sec. 4-5c that our "common sense" ideas about the physical world are for a world in which $\hbar = 0$ and $c = \infty$. In point of fact, the *real* world is a world in which $\hbar$ is finite but very small, while c is finite but very large. Of course, the terms "very small" and "very large" are relative to our customary units of length, mass and time (e.g., the mks units). Now, by simply redefining these units, it is actually possible to obtain a system of units in which

$$\hbar = c = 1 \tag{4-109}$$

The set of units defined by Eqs. (4-109) are called the *natural units*. Recent experiments in the physics of "elementary particles" (e.g., protons, electrons, photons, muons, kaons, neutrinos, etc.) have indicated that the fundamental processes in the universe seem to take place on the scale of the natural units. These processes must therefore be treated within the framework of *relativistic quantum mechanics*, or quantum field theory as it is also called. Even the most cursory discussion of this presumably "ultimate theory" is far beyond the scope of this book. This is partly because the theory is very complicated, and partly because it has not yet been put into a completely consistent and fully understood form.

Broadly speaking, ordinary nonrelativistic quantum mechanics has provided us with a reasonably complete conceptual and quantitative understanding of atomic physics and chemistry; that is, it has been able to explain with remarkable precision the properties of the elements of the periodic table, and thus the properties of all substances which can be formed from these elements. It is hoped that someday relativistic quantum mechanics will be able to afford a similar understanding of the many elementary particles, and of the fundamental mechanisms by which these elementary particles interact with each other to form the various atomic nuclei. It may well be that, in its final form, relativistic quantum theory will be as different from ordinary quantum mechanics, as ordinary quantum mechanics is from classical mechanics. Nevertheless, the nonrelativistic theory of quantum mechanics has and will continue to have a very large range of validity, and it will undoubtedly long stand as one of the most inventive and successful achievements of the human intellect.

Index